Lecture Notes in Computer Science 13410

More information about this subseries at https://link.springer.com/bookseries/8637

Abdelkader Hameurlain ·
A Min Tjoa · Esther Pacitti ·
Zoltan Miklos (Eds.)

Transactions on Large-Scale Data- and Knowledge- Centered Systems LI

Special Issue on Data Management - Principles, Technologies and Applications

 Springer

Editors-in-Chief
Abdelkader Hameurlain
IRIT, Paul Sabatier University
Toulouse, France

A Min Tjoa
IFS, Technical University of Vienna
Vienna, Austria

Guest Editors
Esther Pacitti
University of Montpellier
Montpellier, France

Zoltan Miklos
University of Rennes 1
Rennes, France

ISSN 0302-9743   ISSN 1611-3349 (electronic)
Lecture Notes in Computer Science
ISSN 1869-1994   ISSN 2510-4942 (electronic)
Transactions on Large-Scale Data- and Knowledge-Centered Systems
ISBN 978-3-662-66110-9   ISBN 978-3-662-66111-6 (eBook)
https://doi.org/10.1007/978-3-662-66111-6

This Springer imprint is published by the registered company Springer-Verlag GmbH, DE, part of Springer Nature
The registered company address is: Heidelberger Platz 3, 14197 Berlin, Germany

Preface

This volume contains a selection of fully revised papers from the 37th Conference on Data Management – Principles, Technologies and Applications (BDA 2021). For this special issue, we have selected five articles covering a wide range of timely data management research topics on threats modeling, RDF schema generation, data coverage optimization, data quality, storage on synthetic DNA, and data science.

All authors were invited to prepare and submit journal versions of their contributions which have been fully re-reviewed by the editorial board of this special issue. One of the papers in this special issue is an extended version of an international conference paper published in ISWC 2021.

We would like to take this opportunity to express our sincere thanks to all authors and the editorial board of this special issue for their effort and their valuable contribution in raising the quality of the camera-ready version of the papers.

Finally, we are grateful to the Editors-in-Chief, Abdelkader Hameurlain and A Min Tjoa, for giving us the opportunity to publish this special issue as part of TLDKS Journal series.

July 2022

Esther Pacitti
Zoltan Miklos

Organization

Editors-in-Chief

Abdelkader Hameurlain   Paul Sabatier University, IRIT, France
A Min Tjoa   Technical University of Vienna, IFS, Austria

Guest Editors

Esther Pacitti   University of Montpellier, Inria, and LRIMM, France
Zoltan Miklos   University of Rennes 1, France

Editorial Board

Reza Akbarinia   Inria, France
Dagmar Auer   Johannes Kepler University Linz, Austria
Djamal Benslimane   University of Lyon 1, France
Stéphane Bressan   National University of Singapore, Singapore
Mirel Cosulschi   University of Craiova, Romania
Johann Eder   Alpen Adria University of Klagenfurt, Austria
Anna Formica   National Research Council, Rome, Italy
Shahram Ghandeharizadeh   University of Southern California, USA
Anastasios Gounaris   Aristotle University of Thessaloniki, Greece
Sergio Ilarri   University of Zaragoza, Spain
Petar Jovanovic   Universitat Politècnica de Catalunya (BarcelonaTech), Spain
Aida Kamišalić Latifić   University of Maribor, Slovenia
Dieter Kranzlmüller   Ludwig-Maximilians-Universität München, Germany
Philippe Lamarre   INSA Lyon, France
Lenka Lhotská   Technical University of Prague, Czech Republic
Vladimir Marik   Technical University of Prague, Czech Republic
Jorge Martinez Gil   Software Competence Center Hagenberg, Austria
Franck Morvan   Paul Sabatier University, IRIT, France
Torben Bach Pedersen   Aalborg University, Denmark
Günther Pernul   University of Regensburg, Germany
Viera Rozinajova   Kempelen Institute of Intelligent Technologies, Slovakia
Soror Sahri   LIPADE, Université Paris Cité, France
Joseph Vella   University of Malta, Malta
Shaoyi Yin   Paul Sabatier University, IRIT, France
Feng "George" Yu   Youngstown State University, USA

Editorial Board of Special Issue

Contents

Threats Modeling and Anomaly Detection in the Behaviour of a System - A Review of Some Approaches

Mériem Ghali[1]([✉]), Crystalor Sah[1], Marie Le Guilly[2], and Mohand-Saïd Hacid[2]

[1] Univ Lyon, Université Lyon 1, Villeurbanne, France
{meriem.ghali,crystalor.sah}@etu.univ-lyon1.fr, meriem.ghali@ens-lyon.fr
[2] Univ Lyon, Université Lyon 1, LIRIS (UMR 5205 CNRS), Villeurbanne, France
{marie.le-guilly,mohand-said.hacid}@univ-lyon1.fr

Abstract. With the increase of Big Data, cybersecurity is undergoing massive changes. Because of the vast volume of data, it becomes harder and harder to detect anomalies, and therefore to devise techniques to automatically identify malicious behaviours, even though it is a crucial task. However, Big Data also enables the development of new anomaly detection approaches, based on data analysis and especially machine learning and data mining. With this perspective, it becomes possible to propose solutions that are more flexible and better suited to the new threats that are constantly evolving. In this paper, our objective is to first give a general overview of current approaches used for anomaly detection in the context of cybersecurity, and to implement and test some machine learning techniques for this task, in order to compare their performances. Experiments were carried on the CICIDS2017 dataset, using traditional anomaly detection techniques based on Clustering such as K-Means, EM-Clustering and Classification such as Decision Tree, SVM, Neural Networks.

Keywords: Anomaly detection · Machine learning · Intrusion detection systems · Clustering · Classification · Threat modeling · Attack modeling

1 Introduction

Over the last decade, there has been a huge increase in the volume of shared and collected data. For many systems, this data contains valuable and critical information. For this reason, it is important to design robust and reliable cybersecurity systems, to protect this data from possible attacks or disclosure of sensitive data. These threats are constantly evolving, and it has become complicated to detect, trace and stop those intrusions effectively and in real time. As a result, cybersecurity systems need to be able to adapt to these new challenges.

© Springer-Verlag GmbH Germany, part of Springer Nature 2022
A. Hameurlain et al. (Eds.): *Transactions on Large-Scale Data- and Knowledge-Centered Systems LI*, LNCS 13410, pp. 1–27, 2022.
https://doi.org/10.1007/978-3-662-66111-6_1

Many new solutions have been proposed in order to try to address these new challenges. One important issue is the vulnerability of systems which can be exploited by an attack plotted by a determined opponent. In this context, it is important to be able to detect those attacks, for example by analysing the logs of the system. This is a form of anomaly detection, a well-known but challenging issue, as anomalies tend to be rare and hard to characterise.

In this paper, our first objective is to introduce some cyebrsecurity concepts in order to make the user understand the issues and to raise awareness. For that, we will discuss threat modeling, the existing anomaly detection approaches in the context of cybersecurity, the detection tools as well as the problems related to these techniques, that have been omitted from the literature, namely the attacks and vulnerabilities of machine learning techniques.

Then, we give an overview and discuss about how we implemented some anomaly detection techniques using scikit-learn, especially the ones based on machine learning, to assess their efficiency for suspicious behaviour detection by exploiting the content of log files. We carried out experiments on a dataset containing logs labelled with different attack scenarios and compared the performances of different algorithms to detect those attacks. We consider different metrics in our evaluation of the algorithms. Our contributions can then be summarized as follows:

- An state-of-the-art of recent trends in anomaly and threats detection in cybersecurity systems.
- A review of the anomaly detection approaches and the key components associated.
- A comparison of approaches based on machine learning by considering CICIDS2017 dataset with different type of attacks.

Section 2 presents a state-of-the-art of cybersecurity including brief definitions, modeling techniques, intrusion detection systems, existing detection tools and the related problems. Section 3 presents a review of anomaly detection, the indicators, the way to construct the model and the different techniques which are used. Section 4 presents the experiments. In Sect. 5 we discuss the results of the experiments, a comparative study between the models constructed considering different ways to proceed, and the results reported in other research papers. Finally, we conclude in Sect. 6. The acronyms used in the paper are introduced progressively and listed in Table 1.

Table 1. List of acronyms

Acronym	Description
IDS	Intrusion Detection System
HIDS	Host-based Intrusion Detection System
NIDS	Network based Intrusion Detection System
SIDS	Signature based Intrusion Detection System
AIDS	Anomaly based Intrusion Detection System
ML	Machine Learning
SIEM	Security Information and Event Management
DoS	Denial of Service
DDoS	Distributed Denial-of-service
GMM	Gaussian Mixture Models
SVM	Support Vector Machine
DT	Decision Tree
NN	Neural Network
GNV	Gaussian Naïve Bayes
LP	Label Propagation
DR	Detection Rate
FAR	False Alarm Rate
FP	False Positive
FN	False Negative
TP	True Positive
TN	True Negative
Tr_D	Trained data
Te_D	Tested data

2 Background and Related Work

In order to create a model to detect attacks, we first started to understand the issues of cybersecurity and the existing techniques and tools to identify or stop them. This section represents a state-of-the-art, where we discuss the notion of threat, attack as well as vulnerability, we also talk about two known techniques that help identify a threat. Next, we present Intrusion Detection Systems - IDS, their types and existing tools. Some of these IDS are based on machine learning, but these can be vulnerable or even threatening to the system, so a part of the section deals with the subject in order to make users aware of the stakes. Finally a list of detection tools will be given and presented.

2.1   Threats and Attacks in Cybersecurity

There are several types of dangers that can threaten an organization's systems and networks. These dangers consist in everything that can affect the confidentiality, integrity, or availability of an information system [20]. To better protect critical data efficiently, it is necessary to understand what the system has to be protected from and what kind of vulnerabilities can cause an attack.

First, threats must be identified: they represent the determined opponents able to demonstrate an attack exploiting a vulnerability. These vulnerabilities are also important to analyse and patch, as they are all the security flaws that can be exploited by attackers. Finally, when an attack exploiting a vulnerability happens, it is important to be able to detect it and even better if it can be predicted.

There exists two main kinds of attacks scenarios, both as threatening for a system. With a *passive attack*, the attacker will just observe the messages and copy the data, to get sensitive information. As for an *active attack*, the attacker seeks to modify the content or create fraudulent content.

Because of the wide variety of attacks, they can be hard to anticipate and detect. This is why threat modelling can be a crucial step in identifying an attack, its type and how the opponent operates, this subject is presented in Sect. 2.2. Among the various possible attack scenarios, there exists some well-known patterns [26], such as denial of service, man-in-the-middle, zero-day attacks, etc.

2.2   Threat Modelling

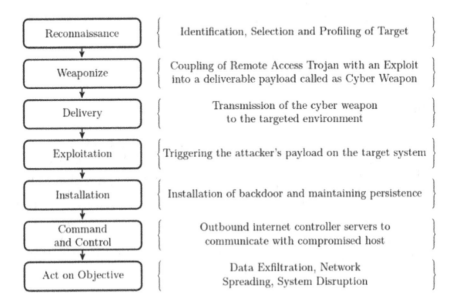

Fig. 1. Phases of cyber kill chain [31]

Threat modelling is a process by which potential threats can be identified, enumerated and mitigations can be prioritized [29]. There are two well-known techniques for vulnerability detection and threat modelling.

First, **STRIDE** [24] is a way to walk through the following categories of threats and determine their potential consequences:

- *Spoofing*: A category of threat that allows a user's identity to be stolen.
- *Tampering*: Falsifying data or code.
- *Repudiation*: Claiming to have not performed an action.
- *Information Disclosure*: Exposing information to someone not authorized to see it.
- *Denial Of Service*: Deny or degrade service to users.
- *Elevation of Privilege*: Gain capabilities without proper authorization.

Second, **Cyber Kill Chain** is a model for incident response teams, digital forensic investigators and malware analysts to work in a chained manner. Inherently understanding, cyber kill chain models and analyses offensive actions of a cyber-attacker [3]. To analyse such attacks, cyber kill chain provides an approach to break down a complicated attack into smaller and easier problems to be analysed and solved.

Cyber Kill chain mainly consists of 7 phases as shown in Fig. 1:

- *Reconnaissance*: The aim of this phase is to gather as much information as possible about the target in order to best carry out the attack.
- *Weaponize*: Using the information collected the opponent creates a penetration plan to gain remote access, the weapon can be seen as a virus or worm.
- *Delivery*: Here the opponent's objective is to transmit the weapon to the target. This can be done either directly using a USB stick or indirectly, requiring interaction with the user who must install the weapon. This stage is very crucial for the opponent because it leaves traces behind [31] and it may be detected and stopped by an IDS.
- *Exploitation*: This part consists of exploiting the vulnerabilities of a system by using the weapon issued to the share before.
- *Installation*: The aim is to install a backdoor.
- *Command and Control*: this step consists in obtaining a privilege elevation and having a root access in order to control the whole machine and to be able to obtain a permanent access.
- *Act on Objectives*: The opponent executes commands according to his motives in order to achieve his goal, his commands may be harmful in the case of an active attack or not in the case of a passive attack.

2.3   Intrusion Detection Systems - IDS

Against the various threats and possible attacks, defence strategies are necessary for preventing data violations or security incidents, monitoring and responding to instructions. To this end, it is interesting to continuously monitor and evaluate

Table 2. IDS tools.

Tool name	Platform	Type of IDS
Snort	Unix, Linux, Windows	NIDS
SolarWind	Windows	NIDS
OSSEC	Unix, Linux, Windows, Mac-OS	HIDS
Security Onion	Linux, Mac-OS	HIDS, NIDS

the daily activities in a network or computer system to detect security risks or threats. Indeed, an Intrusion Detection System (IDS) is a device or software application that monitors a computer network and system for malicious activities or policy violations [12]. There are two main types of Intrusion Detection System:

Host-based Intrusion Detection System (HIDS). The system captures a "picture" of the file set of an entire system and then compares it to a previous picture. If the system finds major discrepancies, such as files that are missing, etc., then it immediately alerts the administrator about it.

Network based Intrusion Detection System (NIDS). The system analyses and monitors network connections to detect all suspicious traffic [20].

In addition to the these two main types of IDS, there are also two main subsets of these IDS types, a third one was proposed to combine the main subsets.

First, **Signature based Intrusion Detection System (SIDS)** takes into account a well-known signature of an attack (string, pattern, rule, etc.), and raises an alarm when it is detected. This technique can be effective in handling a high volume network traffic, and is widely used in the cyber industry. However, it is limited to known attacks only, meaning it will fail to detect new forms of attacks [10].

A second type is **Anomaly based Intrusion Detection System (AIDS)** that analyses the behaviour of a system's network, in order to define what is the *normal* behaviour of a system, and to build a data-driven model able to raise an alarm when a suspicious behaviour is detected [10]. This allows AIDS to detect attacks even if they had not been seen before, therefore overcoming the main issue of SIDS. However, AIDSs also have a tendency to have a high false alarm rate. This issue is therefore an important research question [25]. AIDS are generally based on data mining and machine learning techniques, that have proven efficient in the field of anomaly detection [21,29].

Finally, **Hybrid Intrusion Detection System** combines the strengths from the two previous categories, hybrid IDS use signature-based techniques to detect well-known attacks, and anomaly detection for the new forms of attacks [9,30].

Table 2 lists some of the best IDS tools. Snort is a free open source IDS and intrusion prevention system, SolarWind is mostly used on windows and it consists of collecting NIDS logs, OSSEC is an open source multi-platform that mixes HIDS and log monitoring. Finally, Security Onion is an open source linux distribution for both NIDS and HIDS.

As we aim to build an AIDS model using machine learning, we asked ourselves what potential threats or vulnerabilities we should be aware of in order to avoid having a corrupted model. These are presented in the next section.

2.4   Machine Learning Threats and Vulnerabilities

Machine Learning (*ML*) systems efficiently contribute to the detection of anomalies, but if the system itself is corrupted everything will fall apart, and the results will become meaningless, so in this section we will discuss some risks that can destroy the built system, presented by European Union Agency for Cybersecurity - *ENISA* [2].

- *Evasion:* By exploiting the lack of inputs that defines abnormal behaviour, an optical illusion to the algorithm can be created to cause changes in the output results.
- *Oracle:* This type of attack consists of identifying what type of algorithm is being used, the opponent can cause a series of inputs and observe the outputs and performance of the model. This can be possible if a there is a weak access protection mechanisms for ML model components [2].
- *Poisoning:* known as causative attacks, aim to change the training data labels of ML models by injecting vulnerabilities to modify it's behaviour. This may be caused by undefined indicators of proper functioning, making complex compromise identification [2].
- *Model or data disclosure:* This threat refers to the possibility of leakage of all or part of the model information. This may be caused by Bad practices due to a lack of cybersecurity awareness or existence of unprotected sensitive data on test environments [2].

2.5   Detection Tools

There are several tools available to detect anomalies within a system. These tools automatically and quickly monitor, identify and fix a wide range of security issues, streamlining processes and eliminating the need to perform most repetitive tasks manually. Most modern tools can provide multiple features including automatic detection and blocking of threats and at the same time alerting relevant security teams to further investigate the issue. Here is a list of some detection tools:

- *ManageEngine:* The ManageEngine Event Log Analyser is a SIEM - *Security Information and Event Management* tool that focuses on analysing various logs and extracting various performance and security information from them. The tool, which is ideally a log server, has analytical functions that can identify and report unusual patterns in the logs, such as those resulting from unauthorized access to the organization's IT systems and assets.
- *IBM QRadar:* IBM QRadar SIEM is a detection tool that helps security teams understand threats and prioritize responses. The Qradar takes asset, user, network, cloud, and endpoint data and then correlates it with threat and vulnerability information. After that, it applies advanced scans to detect and track threats as they enter and spread through systems.
- *AlienVault:* AlienVault USM is a comprehensive tool combining threat detection, incident response, and compliance management to provide comprehensive security monitoring and remediation for on-premises and cloud environments. The tool has several security capabilities which also include intrusion detection, vulnerability assessment, asset discovery and inventory, log management, event correlation, email alerts, compliance checks, etc.
- *Sumo Logic:* Sumo Logic is a flexible, cloud-based intelligent security analytics platform that works alone or with other SIEM solutions across multicloud and hybrid environments. The platform uses machine learning to improve threat detection and investigation and can detect and respond to a wide range of security issues in real time. Based on a unified data model, Sumo Logic enables security teams to consolidate security analytics, log management, compliance, and other solutions into one. The solution improves incident response processes in addition to automating various security tasks.
- *McAfee:* McAfee®Network Threat Behaviour Analysis is an integrated component of McAfee Network Security Platform that provides real-time visibility and threat protection of the network infrastructure. By analysing traffic from switches and routers, McAfee Network Threat Behaviour Analysis pinpoints risky behaviour in the network and effectively prevents stealthy attacks. It holistically evaluates network-level threats, identifies the overall behaviour of each network element, and enables instant abstraction of potential anomaly or attack type-including malware, zero-day attacks, botnets, and worms. McAfee Network Threat Behaviour Analysis also houses some of McAfee Network Security Platform's advanced engines, including the real-time emulation engine that identifies malware without signatures.

3   Anomaly Detection for Cybersecurity

Our second objective is to compare the performances of different anomaly detection algorithms in detecting suspicious behaviours. We therefore now give a more comprehensive overview of AIDS, by detailing the main techniques for anomaly detection, and how they are being used in the domain of cybersecurity.

3.1   General Concepts

Anomaly detection aims at detecting an exception and identify an observation that does not comply with an expected outcome. In practice, it is a matter of recognizing which values are problematic among all the data. Anomalies can therefore be defined as patterns in the data that do not comply with the normal behaviour of the system [4].

In the context of cybersecurity, it means that not all anomalies are attacks, but it raises the attention toward data corresponding to an unexpected behaviour which was previously not known. It may or may not be harmful [16]. For this reason, once, for example, an anomaly is detected by an IDS, additional work is required to confirm whether it represents a threat or not. This task is typically done manually by a domain expert.

Another definition is given in the literature for time series data, an anomaly or outlier can be defined as a data point which does not follow the common collective trend or seasonal or cyclic pattern of the entire data and is significantly distinct from the rest of the data. By significant, most data scientists mean statistical significance, which, in other words, means that the statistical properties of the data point is not in alignment with the rest of the series [28].

Since the detection problem, is not easy and not well-defined, there are key components associated with an anomaly detection technique as shown in Fig. 2 that helps us to build the model.

3.2   Characteristics of an Anomaly Detection Problem

The modelling of our detection problem is based on different factors such as the nature of the input data, the availability of the labelled data, the detection techniques and the output or the results of the detection. Besides those factors, we believe that the complexity and the computational time of the chosen detection techniques is very important, especially for real-time detection. This section is a brief review of such factors.

Types of Anomalies. Anomalies can be classified into three categories:

– *Point Anomalies:* This category refers to individual, isolated data points considered anomalous with respect to the rest of data, as it can be seen on Fig. 3(a), *O1* and *O2* represent a point anomaly.

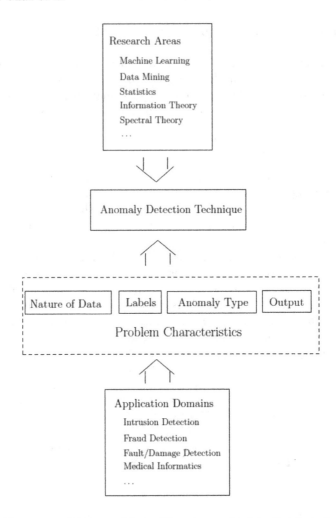

Fig. 2. Key components associated with an anomaly detection technique [16].

- *Contextual Anomalies:* This category refers to single data points, that can be considered anomalous in a specific context, but not otherwise. For example, if one considers weather related data, some points can only be considered anomalies if the seasonal context is taken into account, on Fig. 3(b) t2 is considered anomalous.
- *Collective anomalies:* This category consists in a group of data that is abnormal compared to the rest of the data.

Data Labels. In cybersecurity, one of the main problem of anomaly based techniques is the lack of labelled data and identifying if a data point is normal or anomalous [16]. This labelling is important to characterize the normal behaviour

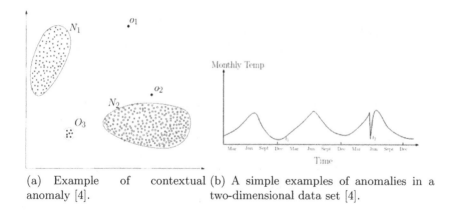

(a) Example of contextual anomaly [4].
(b) A simple examples of anomalies in a two-dimensional data set [4].

Fig. 3. Types of anomalies

of the system, but it is usually done manually by domain experts, which is difficult, tiresome, and prone to errors.

Despite this, labelling normal behaviour is still simpler than characterizing abnormal behaviour. In practice, it is impossible to cover all the possibilities of abnormal behaviour, especially as the latter is often dynamic and there are usually new discoveries of the type of abnormality. It is for these reasons that it is very important to make regular detections and to update the database with new labels. Besides that, the available data has a direct impact on the choice of a method that can be used to create an anomaly detection system and there are three types of techniques:

- *Supervised:* The model will learn from labelled data, containing both normal and abnormal behaviours. The algorithms therefore learn to classify between the two behaviours by training on the labelled data, and can then be used to predict whether or not new incoming data should be considered suspicious.
- *Semi-Supervised:* Only normal labelled data will be needed to build the model. Afterwards, deviations in the test data from that normal model are used to detect anomalies.
- *Unsupervised:* Techniques do not require labelled data. Unsupervised anomaly detection algorithms use only intrinsic information of the data in order to detect instances deviating from the majority of the data.

Output of Anomaly Detection. The outputs produced by anomaly detection techniques can be one of the following two types: [7,16].

- *Scores:* Scoring techniques assign an anomaly score to each point in the test data, depending on the degree to which that point is considered an anomaly. After obtaining the results, a data analyst will study them and decide which data will be considered as an attack.

- *Labels:* Those techniques rather assign a label to each data point - 'normal' or 'anomalous' to each test instance. However, these techniques do not give data analysis experts the right of choice, but it is possible to do so indirectly by choosing the parameters to be taken into account and the algorithms used.

Behavioural Indicators. The detection of anomalies is usually defined by the action of discriminating, in a dataset made up of observations and attributes not known in advance characterizing a target system, those observations that do not correspond to the overall trend represented by the majority.

The problem is how to identify this trend, the execution of a system is defined based on metrics such as response rate to requests or the number of concurrent inputs/outputs in a system. It can also be defined by system resources such as CPU activity or the speed at which a network interface receives network frames, bandwidth, query response times, or error rates during use [22].

3.3   Anomaly Detection Techniques

There exists several possibilities to implement anomaly detection systems. We will now give an overview of those approaches, then we will focus on machine learning techniques that were used in the experiments presented in Sect. 4.

Statistical Based Anomaly Detection. There exists some statistical techniques, based on fitting a statistical model (usually for normal behaviour) to the given data and then apply a statistical inference test to determine if an unseen instance belongs to this model or not. Instances that have a low probability to be generated from the learned model are declared as anomalies [4]. There are two types of statistical techniques:

- *Parametric Techniques:* They assume that normal data are created by parameters and that the anomaly score of a data instance relies on the probability density function related to the parameters. Those techniques can be classified as: Gaussian Model Based, Regression Model Based.
- *Non-parametric Techniques:* With such techniques, the model structure is determined from a given data. It can be categorized as Histogram based.

Classification Based Anomaly Detection. Classification is a process of recognizing, understanding and grouping data into categories. This is done in two stages, a training stage where one uses a pre-categorized dataset to train the model and then a testing stage using a variety of algorithms to classify the future data into categories. Generally, these techniques dominate the supervised algorithms and based on the type of dataset label available the classification can be performed in two ways:

- *One-Class Classification:* Also known as unary classification, the objective is to put all the corresponding data into one category. In our case, the classifier refers to a dataset that describes the normal behaviour of a system, otherwise all the data that belongs to outside the classier will be considered abnormal.
- *Multiple-Class Classification:* It can be also called multinomial classification. This process consists in grouping together instances into **N** classes, the goal is to correctly predict for a point to which class it belongs to. In our case, we consider that a normal system's behaviour can have multiple definitions and if for a given confidence value of an instance no class matches, then it is considered as suspicious.

One of the advantages of classification is the ability to use powerful algorithms, besides that, the test phase is done quickly. It is enough to compare the dataset with the pre-calculated model, it is true that the construction of the model in the training phase can be very slow, but it is not so blocking if it is run offline. What can be challenging in the case of a multi-class classification is to be able to recognize and differentiate all definitions of the normal behaviour.

Clustering Based Anomaly Detection. Generally, clustering algorithms are unsupervised techniques and recently in many papers the semi-supervised clustering was exploited. The purpose of clustering is to divide a data set into different homogeneous clusters, so that the data in each subset share common characteristics. The assumptions made are that for all instances of a given cluster, they should be as similar as possible and reciprocal. Besides that, the difference between the clusters must be clear, but it is not always obvious to satisfy these assumptions when the difference is minimal. According to [4] there are 3 hypotheses for the construction of a clustering algorithm model in anomaly detection:

- *Category 1:* All the points describing the normal behaviour belong to a cluster, points that do not belong to any cluster are considered as abnormal data. These techniques are not optimized to search for anomalies, but focus more on creating clusters; the resulting model from the DBSCAN algorithm is based on this assumption.
- *Category 2:* Normal data are close to the centroid of their nearest neighbouring cluster, otherwise they will be considered as abnormal data. K-means and EM algorithms belong to this category, those techniques can operate on semi-supervised mode. The construction of the resulting model relies on two steps, the first one is the construction of different clusters containing normal data and secondly the use of a distance factor to determine which data will be considered normal or abnormal.
- *Category 3:* The problem with the two previous categories is that if the anomalies themselves form a cluster, the techniques become less successful.

That is why, this category relies on the following assumption: data describing normal behaviour belong to large and dense clusters, while abnormal data belong to different small clusters.

Note that there are other techniques that can are interesting to explore, such as information theoretic anomaly detection and spectral anomaly detection.

Review of Used Techniques

– **Support Vector Machine:** It is a subclass of the classification algorithms, and it can be used to solve discrimination problems, *i.e.*, deciding which class a sample belongs to, or regression problems, *i.e.*, predicting the numerical value of a variable. When compared to results of neural networks it was found that SVM outperformed NN in terms of accuracy but not in terms of false alarm rate as indicated in other research [1,27], the disadvantage of this algorithm is that it requires a lot of memory and CPU time [15]. In our experiments, it was the slowest algorithm. This algorithm only depends on the number of inputs to classify and the number of data learning (n). It therefore has a polynomial complexity. For a very large amount of training data, the computation time explodes. It is why SVMs are more useful for "small" classification problems [13].

– **Neural Network:** It is a set of interconnected nodes designed to imitate the functioning of the human brain. Neural networks supply a very efficient means for detection of net intrusions, as is for example Port scan [19]. In our experiments, we applied the algorithm to a Port scan and DDoS attack.

– **Decision Tree:** Decision trees have several advantages compared to other classification techniques, which make them more suitable for outlier detection. In particular, they have an easily interpretable structure, and they are also less appropriate to the curse of dimensionality [19]. It displays the best result in our experiments. This algorithm has a learning time complexity of $O(nlog(n)* d)$. Regarding the execution time, it is $O(k)$ where k is the depth of the tree.

– **Gaussian Naïve Bayes:** The naïve Bayes model is a heavily simplified Bayesian probability model [14]. The results of the naïve Bayes classifier are often correct, but not always as highlighted with our experiments. It states that the error is a result of three factors: training data noise, bias, and variance [14]. This algorithm is very efficient on large datasets and have a complexity of $O(d*c)$ where d is the query vector's dimension, and c is the total classes.

– **Label Propagation:** Label propagation is a semi-supervised learning algorithm in which labelled and unlabelled data are used to form a similarity or "affinity" matrix W, where w_{ij} represents the similarity between data points i and j [6]. Unlike the result in "Host-based Anomalous Behaviour Detection Using Cluster-Level Markov Networks" [8], the algorithm with our data did not do so well. The time complexity of Label Propagation is nearly linear, each iteration of propagation has complexity $O(m)$, with m being the number of edges.

- **K-Means:** K-Means is an iterative clustering algorithm in which items are processed among a set of clusters until the desired set is reached. Using the K-Means algorithm, we achieve a high degree of similarity among elements in a cluster and dissimilarity in different clusters [18]. However, this technique suffers from some shortcomings. The "wrong choice" of the number of initial groups will decrease the detection of true anomalies and increase the generation of false alarms [5]. If k and d (the dimension) are fixed, the problem can be exactly solved in time $O(n^{dk+1})$, where n is the number of entities to be clustered.

- **DBSCAN:** DBSCAN is an algorithm which allows the partitioning of groups into dense regions separated by less dense regions. Nevertheless, the density-based method has a great influence on the detection capability. It generates suboptimal partitions and considers many instances as noises. Therefore, these problems increase the rate of false positives and decrease the detection rate of real intrusions [5]. This algorithm has a complexity time of $O(nlogn)$. It works well when there is a lot of noise in the dataset and can handle clusters of different shapes and sizes, but if we have a dataset of different densities, the algorithm fails to give an accurate result.

- **EM Clustering:** The Expectation-Maximization algorithm is widely used to estimate the parameters of Gaussian Mixture Models(GMM). GMM is based on the assumption that the data to be clustered are drawn from one of several Gaussian distributions [11]. This algorithm has a linear complexity. The algorithm reads n points. For each point, k probabilities are computed and each of these requires d computations, one per dimension. So each point requires $O(dk)$ work and being n points the total complexity is $O(dkn)$.

4   Experiments

Considering all the available state-of-the-art techniques for anomaly detection in cybersecurity, the objective of these experiments is to evaluate and compare the performances of some of these techniques, in order to better understand their strengths and utility. To this end, we built models based on different techniques; supervised, unsupervised and semi-supervised, and evaluate their performances, that allows us to compare the models' predictions to the ground truth.

The data and code of experimentation are **available** online[1].

4.1   Dataset Description

For the experiments, we used the CICIDS2017 dataset[2] [23], which was designed to evaluate intrusion detection systems. It consists of labelled network flows,

---

[1] https://github.com/gmeriem/Anomaly_Detection_Multiple_Scenarios.
[2] https://www.unb.ca/cic/datasets/ids-2017.html.

contains benign and the most up-to-date common attacks, which resembles the true real-world data (PCAPs). Each line of the dataset represents a flow on the network, and is labelled as either benign, or as an attack. Additionally, there exists different types of attacks, so a line corresponding to an attack is labelled with the corresponding attack's name. The attacks include **Brute Force FTP**, Brute Force SSH, DoS, Heartbleed, Web Attack, Infiltration, Botnet and **DDoS**.

The data capturing period started at 9 a.m., Monday, July 3, 2017 and ended at 5 p.m. on Friday July 7, 2017, for a total of 5 days. Monday is the normal day and only includes the benign traffic. The attacks were executed both morning and afternoon on Tuesday, Wednesday, Thursday and Friday.

4.2   Data Preparation

First, we separated the available data into training and testing sets. In this process, we also cleaned the data, by removing the infinite and null values. The data was then ready to be used in a traditional ML workflow, where the classification models were trained on the training data. Using the data from Monday, that does not contain any attack, we build a *normal behaviour* set, to be used for semi-supervised models.

Training Phase. We created a System profile/Model which represents a normal behaviour based on available labelled data [6], using supervised and semi-supervised classification algorithms available on the scikit-learn library and we measured the execution time of each algorithm.

- Note that to get an anomaly detection in real time we need to modify and refresh the existing profile with timing information. This step is very crucial and sensitive and must be done by a specialist in order to avoid injecting corrupted data that can lead to an attack as explained in Sect. 2.4.
- In the unsupervised case, this phase was not used.

Evaluation Phase. All the models were created using the scikit-learn library [17]. We then used various metrics to assess the performances of the models. In addition to accuracy (Eq. 1), we also used metrics that can better reflect the performances of the models, the detection rate and the false alarm rate [18]. We also used recall, precision and F1-score [6]. These metrics are particularly important for IDS, as most datasets are imbalanced, due to the few number of suspicious data with respect to the rest.

We hereafter recall the definition of these metrics, that are based on the four possible kinds of predictions:

False Positive - FP event signaling IDS to produce an alarm when no attack has taken place [18].

False Negative - FN failure of IDS to detect an actual attack [18].

True Positive - TP legitimate attack which triggers an IDS to produce an alarm [18].

True Negative - TN no attack has taken place and no alarm is raised [18].

$$Accuracy = \frac{TP + TN}{TP + TN + FP + FN} \tag{1}$$

Detection Rate - DR. Detection rate is defined as the number of intrusion instances detected by the system (True Positive) divided by the total number of intrusion instances present in the test dataset, DR is represented in Eq. 2.

$$DR = \frac{TP}{TP + TN + FP + FN} \tag{2}$$

False Alarm Rate - FAR. False Alarm Rate is defined as the number of normal patterns classified as attacks (False Positive) divided by the total number of normal patterns [9], FAR is represented in Eq. 3.

$$FAR = \frac{FP}{FP + TN} \tag{3}$$

We also used score metrics for the clustering, such as Precision, Recall and F1-Score.

Precision. Precision is an indicator that measures the ratio of correctly predicted positive observations to the total predicted positive observations made by the ML model [6], it is presented in Eq. 4.

$$Precision = \frac{TP}{TP + FP} \tag{4}$$

Recall. Recall is the ratio between the true positive predictions and the total of the predictions that the machine should have identified. The maximum value of this indicator for an entity is met when the ML model correctly predicts each occurrence of that entity. [6], it is presented in Eq. 5.

$$Recall = \frac{TP}{TP + FN} \tag{5}$$

F1-score. F1 score can be interpreted as a harmonic mean of the values of the precision and recall indicators, falling within the range [0,1]. [6], it is presented in Eq. 6.

$$F1 - Score = \frac{2 * Recall * Precision}{Recall + Precision} \tag{6}$$

5 Results and Comparative Study

5.1 Scenario 1

Classification Techniques. We tested 4 supervised algorithms and one semi-supervised algorithm on **Brute Force FTP** attacks. The results are shown in Table 3. In this table, it can be seen that the best results are obtained with the decision tree algorithm, that performs extremely well both in terms of accuracy and false alarm rate. Despite its simplicity, this shows the effectiveness of this algorithm for IDS, at least for the CICIDS2017 dataset. It is also important to notice that the only semi-supervised algorithm used performs rather poorly: this indicates that additional work might be necessary to finetune the algorithm on this specific dataset, as it does not seem able to detect suspicious behaviours in the tested configuration. To better understand the difference between the algorithms, we also plotted, on Fig. 4, the predictions given by each algorithm, by projecting the data on two of the dataset's indicators (destination port and flow duration). It can be seen that the decision tree algorithm is better able to detect attacks that are hidden in the middle of normal data. SVM also performs rather correctly, but label propagation fails to detect any suspicious behaviour.

Table 3. Results summary for classification techniques

Parameter	Supervised				Semi-supervised
	SVM	DT	NN	GNB	LP
Accuracy	88,56%	**99,98%**	77.51%	69.16%	44.58%
DR	99.33%	99,98%	99,99%	99.48%	**0.0%**
FAR	24.84%	**0.02%**	50.62%	68.72%	**0.0%**
Time	**45.15 (s)**	1.61 (s)	2.96 (s)	0.12 (s)	8.12 (s)
Train size	143055	143036	143042	143045	**28611**
Test size	143041	143036	143054	143051	**26473**

Clustering Techniques. We used 3 unsupervised clustering algorithms. The results are shown in Tables 4 and 5. For k-mean, several values of k were tested, among which we present three cases, including $k = 1016$ which corresponds to the number of clusters found by EM clustering. From the result tables, it appears that k-mean seems to give the best results, which can indicate that for the CICIDS dataset, the distance-based clustering is more relevant than the others (*e.g.*, density-based clustering). Additionally, we present the clustering predictions in Fig. 5, where all suspicious behaviours are coloured in pink (other colours correspond to normal behaviour). It appears that DBSCAN does not perform really well compared to other algorithms, and that additional work might be necessary to obtain relevant results with DBSCAN and EM clustering on this dataset.

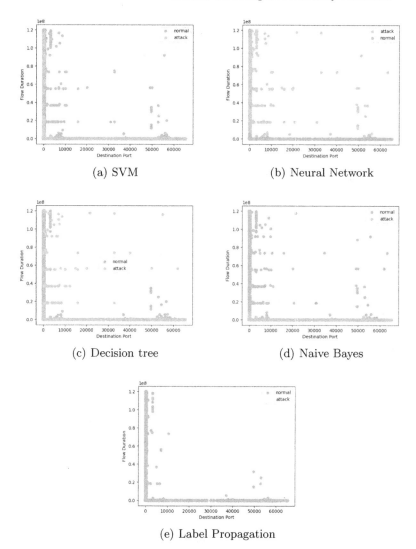

Fig. 4. Comparison of predictions on the CICIDS2017 dataset, between normal behaviour and attacks, for each classification algorithm.

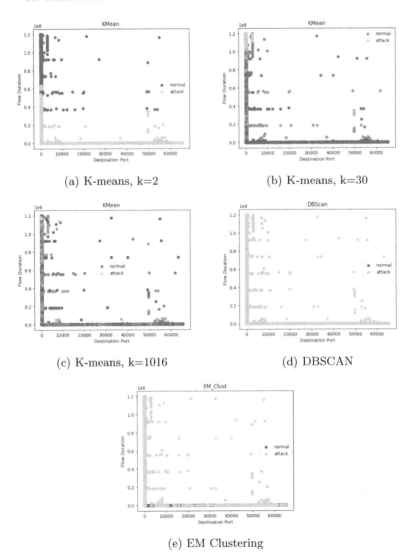

(a) K-means, k=2

(b) K-means, k=30

(c) K-means, k=1016

(d) DBSCAN

(e) EM Clustering

Fig. 5. Comparison of predictions regarding CICIDS2017 dataset, between normal behaviour and attacks, for each clustering algorithm.

Table 4. Results summary for clustering techniques

Algorithms	k	DR	FAR	Accuracy	Test size
K-MEANS	2	**99.89%**	**0.13%**	**60.23%**	**143044**
	30	**99.78%**	0.26%	65.50%	143044
	1016	98.42%	1.91%	**93.84%**	**143044**
DBSCAN	1016	**93.02%**	7.95%	**51.99%**	**143044**
EM-Clustering		**0.92%**	55.38%	27.84%	143044

Table 5. Confusion scores for clustering techniques

	KMean k = 1016		DBSCAN		EM_Clustering	
	N	A	N	A	N	A
Recall	88%	**98%**	01%	93%	62%	1%
F1-Score	93%	95%	02%	**68%**	**43%**	1%
Precision	98%	**91%**	12%	**54%**	**33%**	3%
Support	**63776**	**79227**	63649	79414	63553	79489

5.2 Scenario 2

Classification Techniques. In this scenario, we selected the most important features as mentioned above in Sect. 3.2, and it was tested on 4 supervised techniques. As shown in the table below Table 6, DT achieves the best results, but the NN algorithms describes all points as an attack and increases the score of the FAR as shown in Fig. 6.

This can be explained with the fact that the points that have been detected have very few characters in common and one has to be more precise with the parameters in order to distinguish the difference. By comparing the results with those obtained in the previous scenario and the resulting Table 3, SVM has a better CPU time and the results are not that different.

Table 6. Results summary for classification techniques, scenario 2.

Parameter	Supervised			
	SVM	DT	NN	GNB
Accuracy	83.20%	**99,97%**	**57.51%**	64.16%
DR	97.05%	99,97%	100%	98.97%
FAR	34.16%	**0.03%**	94.88%	79.11%
Time	**26.06 (s)**	0.48 (s)	2.83 (s)	0.03 (s)
Train size	143055	143036	143042	143045
Test size	143041	143036	143054	143051

Clustering Techniques. For the clustering techniques, we can see a slight improvement for the K-Means regarding the DR and the accuracy with k = 1016, but globally there is not that much difference. We can also note that the EM-Clusturing has poor performance regarding the detection of anomalies (Tables 7 and 8).

Table 7. Results summary for clustering techniques, scenario 2.

Algorithms	k	DR	FAR	Accuracy	Test size
K-MEANS	1016	99.09%	1.11%	**95.15%**	**143031**
DBSCAN	1016	**93.29%**	7.75%	**53.23%**	**143065**
EM-Clustering		**0.0%**	55.69%	44.27%	143036

Table 8. Confusion scores for clustering techniques, scenario 2.

	KMean k = 1016		DBSCAN		EM_Clustering	
	N	A	N	A	N	A
Recall	90%	**99%**	03%	93%	100%	0%
F1-Score	94%	96%	05%	**69%**	**61%**	0%
Precision	99%	**95%**	26%	**55%**	**44%**	0%
Support	**63520**	**79511**	63492	79573	63366	79670

5.3 Scenario 3

Classification Techniques. In this scenario, the detection of another attack, **DDoS,** has been exploited and it is more similar to scenario 1 which means that all features are considered by default. In this scenario, DT also outperformed the other techniques, what may also be remarkable is the execution time of the SVM model when comparing the result with the previous Table 3 but contrary to Table 6, the results are better.

In addition to this, as shown in Fig. 8, the model resulting from the NN algorithm and the GNB algorithm reacted badly and considered the majority of the points as anomalies.

Clustering Techniques. For the clustering techniques we can see that DBScan have for this data a good DR, but this is not very relevant since it does not have very good accuracy. For the K-Means it still has good performance (Fig. 9 and Tables 9, 10, 11).

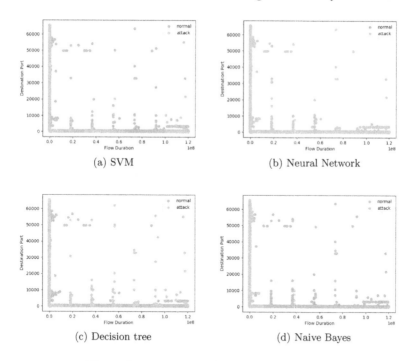

(a) SVM

(b) Neural Network

(c) Decision tree

(d) Naive Bayes

Fig. 6. Scenario 2 of Classification prediction.

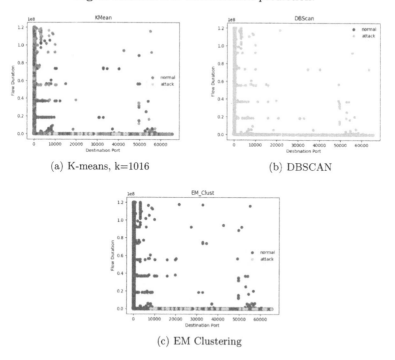

(a) K-means, k=1016

(b) DBSCAN

(c) EM Clustering

Fig. 7. Scenario 2 of Clustering prediction.

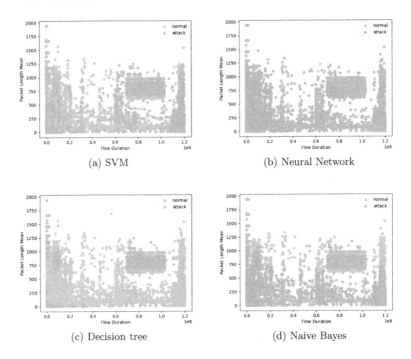

(a) SVM

(b) Neural Network

(c) Decision tree

(d) Naive Bayes

Fig. 8. Scenario 3 of Classification prediction.

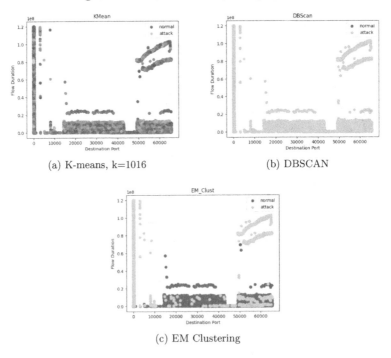

(a) K-means, k=1016

(b) DBSCAN

(c) EM Clustering

Fig. 9. Scenario 3 of Clustering prediction.

Table 9. Results summary for classification techniques, scenario 3.

Parameter	Supervised			
	SVM	DT	NN	GNB
Accuracy	**95.54%**	**99,98%**	74.16%	78.98%
DR	98.21%	99,99%	**99,99%**	99.88%
FAR	7.97%	**0.02%**	**59.77%**	48.34%
Time	**33.53 (s)**	0.811 (s)	2.37 (s)	0.05 (s)
Train size	112855	112852	112857	112851
Test size	112856	112859	112854	112860

Table 10. Results summary for clustering techniques, scenario 3.

Algorithms	k	DR	FAR	Accuracy	Test size
K-MEANS	1016	98.01%	2.53%	**96.647%**	**112859**
DBSCAN	1016	**100%**	0.0%	**56.92%**	**112849**
EM-Clustering		**53.10%**	38.09%	60.30%	112853

5.4 Discussion

In this ongoing study on machine learning algorithms for IDS, we implemented and tested several techniques of the state-of-the-art on a well-known dataset, in order to discriminate between normal and suspicious behaviour. Based on our initial results, it appears that for a given dataset, there is a great variability in the performances of the algorithms, leading to conclude that it is necessary to identify the most suited one for each new dataset. Additionally, it would be interesting to better study the impact of the parameters of each algorithm on its performances.

Table 11. Confusion scores for clustering techniques, scenario 3.

	KMean k = 1016		DBSCAN		EM_Clustering	
	N	A	N	A	N	A
Recall	95%	**98%**	01%	**100%**	70%	53%
F1-Score	96%	97%	01%	72%	60%	60%
Precision	97%	96%	**100 %**	57%	53%	70%
Support	**48775**	**64084**	48926	63923	48809	64044

6 Conclusion and Future Work

In this work, we reviewed and tested algorithms to effectively detect an attack before it gets into the system. Even if the threat succeeds in getting into the system, we described approaches based on supervised, semi-supervised and unsupervised algorithms to detect anomalies in order to unmask an attack and stop it, which can allow administrators to take measures to prevent such attacks to happen again in the future.

We tested those algorithms by using a well-known dataset, and we presented the results related to the behaviour of the algorithms (*e.g.*, performances). The results show that the Decision Tree and K-Means algorithms are the best ones in terms of detection and accuracy with a low false alarm rate, for the considered CICIDS2017 dataset. We believe that an efficient detection of anomalies remains a relevant issue to be studied in depth. For future work, it would be interesting to build detection models based on a combination of algorithms.

References

1. Agrawal, S., Agrawal, J.: Survey on anomaly detection using data mining techniques. Procedia Comput. Sci. **60**, 708–713 (2015)
2. Malatras, A., Ioannis Agrafiotis, M.A.: Securing machine learning algorithms. European Union Agency for Cybersecurity - ENISA, December 2021
3. Bayuk, J., et al.: Malware risks and mitigation report, vol. 21, p. 139. BITS Financial Services Roundtable, Washington, DC (2011)
4. Chandola, V., Banerjee, A., Kumar, V.: Anomaly detection: a survey. ACM Comput. Surv. (CSUR) **41**(3), 1–58 (2009)
5. Cuxac, P., Lamirel, J.C.: Clustering incrémental et méthodes de détection de nouveauté: application à l'analyse intelligente d'informations évoluant au cours du temps (2011)
6. Georgescu, T.M.: Natural language processing model for automatic analysis of cybersecurity-related documents. Symmetry **12**(3), 354 (2020)
7. Goldstein, M., Uchida, S.: A comparative evaluation of unsupervised anomaly detection algorithms for multivariate data. PLoS ONE **11**(4), e0152173 (2016)
8. Ingram, J.B., Chiang, K., Mustafa, A., Solaimani, M., Sahs, J., Khan, L.: Host-based anomalous behavior detection using cluster-level Markov networks. Technical report, Sandia National Lab. (SNL-NM), Albuquerque, NM (United States); Sandia (2013)
9. Kim, G., Lee, S., Kim, S.: A novel hybrid intrusion detection method integrating anomaly detection with misuse detection. Expert Syst. Appl. **41**(4), 1690–1700 (2014)
10. Liao, H.J., Lin, C.H.R., Lin, Y.C., Tung, K.Y.: Intrusion detection system: a comprehensive review. J. Netw. Comput. Appl. **36**(1), 16–24 (2013)
11. Lu, W., Tong, H.: Detecting network anomalies using CUSUM and EM clustering. In: Cai, Z., Li, Z., Kang, Z., Liu, Y. (eds.) ISICA 2009. LNCS, vol. 5821, pp. 297–308. Springer, Heidelberg (2009). https://doi.org/10.1007/978-3-642-04843-2_32
12. Milenkoski, A., Vieira, M., Kounev, S., Avritzer, A., Payne, B.D.: Evaluating computer intrusion detection systems: a survey of common practices. ACM Comput. Surv. (CSUR) **48**(1), 1–41 (2015)
13. Mohamadally Hasan, F.B.: SVM: Machines à vecteurs de support séparateurs à vastes marges. BD Web ISTY3 **21**, 14–15 (2006)
14. Mukherjee, S., Sharma, N.: Intrusion detection using Naive Bayes classifier with feature reduction. Procedia Technol. **4**, 119–128 (2012)

15. Omar, S., Ngadi, A., Jebur, H.H.: Machine learning techniques for anomaly detection: an overview. Int. J. Comput. Appl. **79**(2) (2013)
16. Parmar, J.D., Patel, J.T.: Anomaly detection in data mining: a review. Int. J. **7**(4), 32–40 (2017)
17. Pedregosa, F., et al.: Scikit-learn: machine learning in Python. J. Mach. Learn. Res. **12**, 2825–2830 (2011)
18. Ranjan, R., Sahoo, G.: A new clustering approach for anomaly intrusion detection. arXiv preprint arXiv:1404.2772 (2014)
19. Reif, M., Goldstein, M., Stahl, A., Breuel, T.M.: Anomaly detection by combining decision trees and parametric densities. In: 2008 19th International Conference on Pattern Recognition, pp. 1–4. IEEE (2008)
20. Sarker, I.H., Kayes, A.S.M., Badsha, S., Alqahtani, H., Watters, P., Ng, A.: Cybersecurity data science: an overview from machine learning perspective. J. Big Data **7**(1), 1–29 (2020). https://doi.org/10.1186/s40537-020-00318-5
21. Sarker, I.H., Kayes, A.S.M., Watters, P.: Effectiveness analysis of machine learning classification models for predicting personalized context-aware smartphone usage. J. Big Data **6**(1), 1–28 (2019). https://doi.org/10.1186/s40537-019-0219-y
22. Sauvanaud, C.: Monitoring et détection d'anomalie par apprentissage dans les infrastructures virtualisées. Ph.D. thesis, Toulouse, INSA (2016)
23. Sharafaldin, I., Lashkari, A.H., Ghorbani, A.A.: Toward generating a new intrusion detection dataset and intrusion traffic characterization. In: ICISSp, pp. 108–116 (2018)
24. Shostack, A.: Threat Modeling: Designing for Security. Wiley, Hoboken (2014)
25. Sommer, R., Paxson, V.: Outside the closed world: on using machine learning for network intrusion detection. In: 2010 IEEE Symposium on Security and Privacy, pp. 305–316. IEEE (2010)
26. Sun, N., Zhang, J., Rimba, P., Gao, S., Zhang, L.Y., Xiang, Y.: Data-driven cybersecurity incident prediction: a survey. IEEE Commun. Surv. Tutor. **21**(2), 1744–1772 (2018)
27. Tang, H., Cao, Z.: Machine learning-based intrusion detection algorithms. J. Comput. Inf. Syst. **5**(6), 1825–1831 (2009)
28. Teng, M.: Anomaly detection on time series. In: 2010 IEEE International Conference on Progress in Informatics and Computing, vol. 1, pp. 603–608. IEEE (2010)
29. Tsai, C.F., Hsu, Y.F., Lin, C.Y., Lin, W.Y.: Intrusion detection by machine learning: a review. Expert Syst. Appl. **36**(10), 11994–12000 (2009)
30. Viegas, E., Santin, A.O., Franca, A., Jasinski, R., Pedroni, V.A., Oliveira, L.S.: Towards an energy-efficient anomaly-based intrusion detection engine for embedded systems. IEEE Trans. Comput. **66**(1), 163–177 (2016)
31. Yadav, T., Rao, A.M.: Technical aspects of cyber kill chain. In: Abawajy, J.H., Mukherjea, S., Thampi, S.M., Ruiz-Martínez, A. (eds.) SSCC 2015. CCIS, vol. 536, pp. 438–452. Springer, Cham (2015). https://doi.org/10.1007/978-3-319-22915-7_40

Incremental Schema Generation for Large and Evolving RDF Sources

Redouane Bouhamoum[(⊠)], Zoubida Kedad, and Stéphane Lopes

DAVID Laboratory, University of Versailles Saint-Quentin-en-Yvelines,
Versailles, France
{redouane.bouhamoum,zoubida.kedad,stephane.lopes}@uvsq.fr

Abstract. The lack of a descriptive schema for an RDF dataset has motivated several research works addressing the problem of automatic schema discovery. The goal of these approaches is to provide the underlying structural schema of a given RDF dataset, either from the existing instances, or using some schema-related declarations if provided. However, as the instances in the RDF dataset evolve, the generated schema may become inconsistent with the dataset. It is therefore necessary to incrementally update the existing schema according to the changes occurring in the dataset over time.

In this paper, we propose a schema discovery approach for massive RDF datasets which incrementally deals with both the insertion and the deletion of entities. It is based on a scalable and incremental density-based clustering algorithm which propagates the changes occurring in the dataset into the clusters corresponding to the classes of the schema. Our approach is implemented using big data technologies to scale-up to massive data, while providing a high quality clustering result. We present some experiments which demonstrate the efficiency of our proposal on both synthetic and real datasets.

Keywords: Incremental schema discovery · Schema evolution · RDF data · Big data · Clustering

1 Introduction

The web of data represents a huge information space consisting of an increasing number of interlinked datasets described using the Resource Description Framework (RDF). RDF is a standard model for data interchange on the Web and a recommendation of the W3C for representing and publishing datasets. One important feature of such datasets is that they contain both the data and the schema describing the data. However, these schema-related declarations are not mandatory, and are not always provided. As a consequence, the schema may be incomplete or missing. Furthermore, even if the schema is provided, it is not a constraint on the data: resources of the same type may be described by property sets which are different from those specified in the schema. In the context of web

© Springer-Verlag GmbH Germany, part of Springer Nature 2022
A. Hameurlain et al. (Eds.): *Transactions on Large-Scale Data- and Knowledge-Centered Systems LI*, LNCS 13410, pp. 28–63, 2022.
https://doi.org/10.1007/978-3-662-66111-6_2

data, the generated schema is viewed as a guide easing the exploitation of the RDF dataset, and not as a structural constraint over the data.

The lack of schema offers a high flexibility while creating interlinked datasets, but can also limit their use. Indeed, it is not easy to query or explore a dataset without any knowledge of its resources, classes or properties. For instance, writing a query in SPARQL, the standard query language for RDF, requires some knowledge about property names, entities or resources in the dataset.

The exploitation of an RDF dataset would be easier with a schema describing the data. The schema is useful for various data processing and data management tasks such as improving the indexing of online content, processing queries more efficiently, giving more relevant responses, and increasing interoperability between systems. Generating a schema from the entities of RDF datasets is thus an important task to ease their exploitation. We have proposed in previous work a schema discovery approach suitable for very large datasets [5]. To this end, we have introduced a scalable density-based clustering algorithm specifically designed for schema discovery in large RDF datasets. Using the proposed clustering algorithm, we explore instance-level data in an RDF dataset in order to infer a schema composed of classes and their properties. This approach allows performing fast density-based clustering on large datasets and provides a good quality schema.

However, RDF datasets are subject to frequent evolutions over time, new instances may be inserted and existing ones may be deleted. For example, between the version 3.5 and the version 3.9 of DBpedia, the number of triples having the class *Person* as their subject has been multiplied by 45 [29]. Due to this evolution, the ability to update the schema incrementally has emerged as a new challenge. Existing approaches require the availability of the whole dataset before running the algorithm [8, 23, 33], or introduce the notion of fictive entity associated to each cluster to classify the new inserted elements in the existing clusters [24] without creating, deleting or merging clusters. Indeed, in many applications where the time factor is important and where the datasets are big in size, executing the schema discovery process on the whole dataset to update the schema after each evolution is a very costly process, especially if this latter relies on a clustering algorithm. This highlights the need for approaches that can incrementally update a schema describing an RDF dataset, without processing the whole dataset after each update.

In this work, we introduce an incremental schema discovery approach for large RDF data. Our contribution is an incremental density-based clustering algorithm for building and updating the clusters that represent the classes of the schema. Our proposal incrementally updates the classes describing an RDF dataset in order to keep the schema consistent with the evolution of the data and ensures providing the same result as if the clustering algorithm has been executed on the whole dataset in one go. In addition, the incremental clustering process is parallelized to be efficient on large datasets. The source code of the implementation of our algorithm, based on the distributed processing framework Apache Spark [34] is available online[1].

---

[1] https://github.com/BOUHAMOUM/incremental_sc_dbscan.git.

The present paper is an extended version of the work presented in [6], where we have introduced an incremental approach for schema discovery dealing with the insertion of sets of entities. Beside providing a more detailed description of our approach, we present in this paper an extension of our proposal in order to deal with the deletion of sets of entities as well as insertions, and we provide the corresponding experimental evaluations to show the efficiency of our algorithms. Dealing with deletion is equally important as dealing with insertions as they may both lead to a schema which is inconsistent with the dataset it describes; furthermore, the size of the deleted set of entities can be massive, requiring scalable solutions.

The rest of the paper is organized as follows. Section 2 presents the problem addressed in this paper and provides some preliminary notations. Section 3 presents our proposal for schema evolution when dealing with insertions, and Sect. 4 details how our approach deals with deletions. Experimental results are presented in Sect. 5, and Sect. 6 discusses the related works. Finally, Sect. 7 concludes the paper and presents some future works.

2 Problem Statement

An RDF *dataset* is a set of RDF(S)/OWL triples $\mathcal{D} \subseteq (\mathcal{R} \cup \mathcal{B}) \times \mathcal{P} \times (\mathcal{R} \cup \mathcal{B} \cup \mathcal{L})$, where $\mathcal{R}$, $\mathcal{B}$, $\mathcal{P}$ and $\mathcal{L}$ represent resources, blank nodes (anonymous resources), properties and literals respectively. In such dataset, an *entity e* is either a resource or a blank node, i.e. $e \in \mathcal{R} \cup \mathcal{B}$. We denote by D the set of entities of the dataset $\mathcal{D}$.

We define a function, denoted $\overline{}$, which returns the set of properties of an entity: $\overline{e} = \{p \in \mathcal{P} \mid \langle e, p, o \rangle \in D\}$. This function can be extended to represent the set of properties of a set of entities $E \subseteq D$: $\overline{E} = \bigcup_{e \in E} \overline{e}$. The dataset D is described by the schema S, defined as follows.

Definition 1. *A schema S describing a dataset D is composed of a set of classes $\{C_1, \ldots, C_n\}$, where each C_i is described by the set of properties $\overline{C_i} = \{p_1^i, \ldots, p_{m_i}^i\}$.*

We define below the updates of a dataset D that are considered in our approach.

Definition 2. *A dataset D can be updated by inserting a set of entities denoted by Δ_D^+ or by deleting a set of entities denoted by Δ_D^-.*

The updates of D with Δ_D^+ or Δ_D^- may result in S to become incoherent with the new dataset $D \cup \Delta_D^+$ or $D \setminus \Delta_D^-$. In other words, S is no longer a good description for the updated dataset.

To deal with this problem, we make the following assumptions:

1. The datasets D, Δ_D^+ and Δ_D^- can be massive. In the big data era, RDF data might be produced in high volumes just like all the other kinds of data.

2. The schema S describing the dataset D has been generated using the principles of the density based clustering algorithm (DBSCAN) [11] which has been used in existing works for schema discovery on RDF data and provided good quality results [5,24]. DBSCAN requires two parameters: the similarity threshold ϵ and the density threshold $minPts$.

3. The entities of the dataset are compared using the *Jaccard index* which is defined as the size of the intersection of the property sets divided by the size of their union [19]:

$$\forall e_i, e_j \in D, J(e_i, e_j) = \frac{|\overline{e_i} \cap \overline{e_j}|}{|\overline{e_i} \cup \overline{e_j}|}$$

Two entities for which the value of the Jaccard index is above the similarity threshold ϵ are neighbors. Note that the similarity between entities could be evaluated using other similarity indices which measure the similarity between finite sets such as the *Sørensen-Dice* [14] or *Overlap* [1] indexes. In our work, we use the *Jaccard index* which has been used in several schema discovery approaches [8,23,24], leading to a good quality schema.

4. For each entity $e \in D$, its neighborhood is known. We assume that the neighborhood of the clustered entities has been computed and saved during previous extractions of the schema.

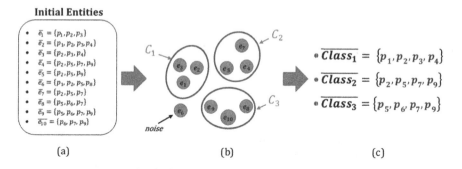

(a)   (b)   (c)

Fig. 1. A set of entities and the corresponding schema

Figure 1 presents a set of entities (Fig. 1a) grouped into three clusters (Fig. 1b) using DBSCAN. The similarity threshold ϵ is set to 0.7 and the density threshold $minPts$ to 2. The resulting clusters represent the classes of the schema (Fig. 1c).

In this work, our aim is to update the schema S considering updates in Δ_D^+ or Δ_D^-. In order to update this schema, we have to modify the set of classes impacted by the insertion or the deletion of entities. The resulting schema after the propagation of updates in the set of existing classes is a descriptive schema which represents the updated dataset.

In the following section, we first introduce our proposal for updating the schema with respect to the insertion of a set of new entities. Dealing with the deletion of a set of entities is detailed in Sect. 4.

3   Schema Evolution After Entity Insertions

In this section, we describe an incremental distributed density-based clustering algorithm to extract a schema from large RDF datasets that evolve over time. It allows to keep the schema coherent with the dataset when new entities are added. In order to efficiently manage incrementally growing big datasets, the clustering is restricted to new entities and their neighborhood within the old entities. Clustering the new entities and updating the clusters within their neighborhoods ensures providing the same result as executing DBSCAN on the global dataset [10].

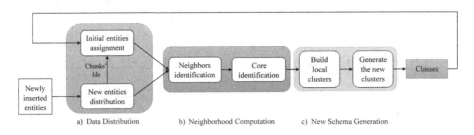

Fig. 2. Overview of our incremental schema discovery approach for dealing with insertions

Our approach is composed of three main steps parallelized and implemented using a big data technology. Figure 2 illustrates the different steps.

First, data are split into subsets, called *chunks*, in order to distribute the entities over the different processes. The chunks contain entities sharing some properties and which are likely to be similar. The distribution is limited to new entities from Δ_D^+ and only old entities from D that are likely to be similar to new ones. As presented in Fig. 2a, the new entities are distributed, then the identifiers of the created chunks are used for the assignment of old entities contained in the initial dataset. This way, all the entities that could be similar to new ones, whether in D or Δ_D^+, are grouped in at least one chunk and consequently all the relevant comparisons will be performed during the clustering step.

Second, in parallel on each node, the neighborhood for each new entity is computed. Then the list of their neighbors which could be distributed over different chunks are merged into a single list. At the end of this step, entities having dense neighborhoods, called *core entities*, are identified. This step is presented in Fig. 2b.

Finally, based on the neighborhood of the new entities, the set of clusters is built locally in each chunk: (i) old clusters may be updated by adding new elements, (ii) some old clusters may be merged, and (iii) new clusters may be created to represent new entities. The clusters produced within each chunk are then merged to generate the new clusters which represent the classes of the new schema as illustrated in Fig. 2c.

We have implemented our algorithm using Spark [34], a big data technology enabling a fast distributed execution of the approach and allowing to manage massive datasets. We detail in the following sections the different steps of our proposal.

3.1 Data Distribution Principle for Neighborhood Computation

Computing the neighborhood of the new entities may require a very high number of comparisons. We propose to distribute these new entities according to the distribution principle introduced in [5], where the entities of the dataset are split into different subsets according to their properties. The comparison of entities is performed within each chunk in parallel, thus speeding-up the clustering process.

The intuition behind our distribution method is to group entities sharing some properties into chunks to ensure that all the pairs of similar entities will be detected. Indeed, according to the similarity index, two entities are similar if they share a number of properties higher than a given threshold. Thus, entities that could be similar are grouped together in at least one chunk, and will be compared during the computation of their neighborhood. If two entities are not grouped in any of the resulting chunks, this means that they are not similar. This distribution principle allows to skip meaningless comparisons as the similarity between entities in different chunks is not evaluated.

In this section, we first describe how to split the new dataset into chunks, then we show how to assign existing entities to the created chunks by identifying the ones that could be similar to one of the newly inserted entities.

Distributing New Entities over Chunks. In our incremental algorithm, we first distribute the new entities according to the properties describing them. An entity is distributed over several chunks according to its properties in order to ensure that it will be compared to all its neighbors.

To optimise the distribution of entities in our approach, we do not consider all the properties of the entities. We therefore limit the duplication of entities in the different chunks and reduce the cost of the comparison process by skipping useless comparisons.

Recall that, as stated in Sect. 2, the similarity between entities is computed using the Jaccard index and that two entities for which the value of the Jaccard index is above the similarity threshold ϵ are neighbors. The following proposition expresses that if the properties of two entities differ to a certain extent, these entities cannot be similar.

Proposition 1. *Let e_1 and e_2 be two entities and ϵ the similarity threshold. If $|\overline{e_1} \setminus \overline{e_2}| \geq |\overline{e_1}| - \lceil \epsilon \times |\overline{e_1}| \rceil + 1$ then e_1 and e_2 can not be similar.*

Proof. Suppose that $|\overline{e_1} \setminus \overline{e_2}| \geq |\overline{e_1}| - \lceil \epsilon \times |\overline{e_1}| \rceil + 1$. We have $|\overline{e_1} \setminus \overline{e_2}| = |\overline{e_1}| - |\overline{e_1} \cap \overline{e_2}|$. Thus, $|\overline{e_1}| - |\overline{e_1} \cap \overline{e_2}| \geq |\overline{e_1}| - \lceil \epsilon \times |\overline{e_1}| \rceil + 1$. By eliminating $|\overline{e_1}|$ on both sides, we obtain $|\overline{e_1} \cap \overline{e_2}| \leq \lceil \epsilon \times |\overline{e_1}| \rceil - 1$ which implies that $|\overline{e_1} \cap \overline{e_2}| < \lceil \epsilon \times |\overline{e_1}| \rceil$. According to the definition of the Jaccard similarity index, this formula implies that e_1 and e_2 can not be similar.

From this proposition, the *dissimilarity threshold* for an entity e is defined as follows:

Definition 3. *The* dissimilarity threshold *for an entity e is the number $dt(e) = |\overline{e}| - \lceil \epsilon \times |\overline{e}| \rceil + 1$.*

This threshold represents the number of properties to consider in order to decide whether this entity could be similar to any other one. It allows to reduce the number of entities to be considered when searching for the neighborhood of a given entity. Note that the *dissimilarity threshold* as defined in our work is based on the Jaccard similarity index. Using another index would require to propose another definition of this threshold based on this index.

However, properties can not be selected randomly, otherwise, this will prevent similar entities to be grouped in the same chunks and compared later. For example, let us consider the similar entities e_2 and e_3 where $\overline{e_2} = \{p_1, p_2, p_3, p_4\}$ and $\overline{e_3} = \{p_2, p_3, p_4\}$. Assuming that the similarity threshold is $\epsilon = 0.7$, and considering the dissimilarity threshold, the entity e_2 can be assigned to $[p_1]$, $[p_2]$ and e_3 only to $[p_3]$. e_2 and e_3 are not grouped in a same chunk even though they are similar. We can see that randomly assigning these entities does not guarantee that they are compared even if they are similar. This problem can be solved by defining a total order on the properties and selecting the properties according to this order. In this example, if we consider an order on the properties during the assignment, the entity e_3 would be assigned to $[p_2]$ instead of $[p_3]$. Therefore, e_2 and e_3 would be grouped in the chunk $[p_2]$ and compared during the computation of their neighborhood. In our work, we propose to order the properties according to their selectivity. This will lead to smaller chunks and therefore the clustering will be more efficient as a higher number of meaningless comparisons will be avoided.

Definition 4. *Let $<_{\mathcal{P}}$ be a total order on the properties describing a dataset, e an entity such that $\overline{e} = \{p_1, p_2, \ldots, p_n\}$ and $p_i <_{\mathcal{P}} p_{i+1}$ for $1 \leq i < n$. The comparison set of e denoted by $cs(e)$ is the set of properties $\{p_1, p_2, \ldots, p_{dt(e)}\}$.*

We will now introduce the definition of a *chunk*.

Definition 5. *A chunk for a property $p \in \mathcal{P}$ denoted by $[p]$ is a subset of entities having the property p in their comparison set: $[p] = \{e \mid p \in cs(e)\}$.*

Proposition 2. *By comparing only entities inside chunks, all the comparisons required for the clustering will be performed at least once.*

Proof. We have to show that if two entities are similar, then there is at least one chunk where both entities have been assigned. Let e_1 and e_2 be two similar entities. We have $\frac{|\overline{e_1} \cap \overline{e_2}|}{|\overline{e_1} \cup \overline{e_2}|} \geq \epsilon$. Thus, $|\overline{e_1} \cap \overline{e_2}| \geq \epsilon \times |\overline{e_1} \cup \overline{e_2}|$ which implies that $|\overline{e_1} \cap \overline{e_2}| \geq \lceil \epsilon \times |\overline{e_1}| \rceil$ and $|\overline{e_1} \cap \overline{e_2}| \geq \lceil \epsilon \times |\overline{e_2}| \rceil$. This implies that $|\overline{e_1}| - |\overline{e_1} \cap \overline{e_2}| \leq |\overline{e_1}| - \lceil \epsilon \times |\overline{e_1}| \rceil$. As $|\overline{e_1} \setminus \overline{e_2}| = |\overline{e_1}| - |\overline{e_1} \cap \overline{e_2}|$, we obtain $|\overline{e_1} \setminus \overline{e_2}| \leq |\overline{e_1}| - \lceil \epsilon \times |\overline{e_1}| \rceil$. As $|cs(e_1)| = dt(e_1) > |\overline{e_1}| - \lceil \epsilon \times |\overline{e_1}| \rceil$, we have $cs(e_1) \cap \overline{e_2} \neq \emptyset$.

We can show likewise that $cs(e_2) \cap \overline{e_1} \neq \emptyset$. Consequently, $cs(e_1)$ and $cs(e_2)$ contain both an element of $\overline{e_1} \cap \overline{e_2}$.

If there is a total order on the set of properties, we can choose the infimum of $\overline{e_1} \cap \overline{e_2}$ for $cs(e_1)$ and $cs(e_2)$. In this case, $cs(e_1) \cap cs(e_2) \neq \emptyset$. This means that at least one chunk will contain both e_1 and e_2.

Example 1. In our example, we consider the dataset D introduced in Fig. 1 and the set of new entities Δ_D^+ presented in Fig. 3. With respect to the selectivity of the properties describing the new entities, the order is $p_4 <_P p_7 <_P p_3 <_P p_8 <_P p_{12} <_P p_1 <_P p_5 <_P p_9 <_P p_{10} <_P p_{11}$. The distribution of the new entities within Δ_D^+ into chunks provides the result presented in Fig. 3.

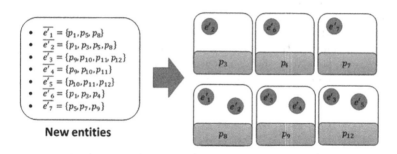

Fig. 3. Chunks generated from the distribution of the new entities in Δ_D^+

For example, the entity e_3' is assigned to the chunks $[p_9]$ and $[p_{12}]$ since its dissimilarity threshold is equal to $dt(e_3') = 4 - \lceil 0.7 \times 4 \rceil + 1 = 2$, and the two most selective properties describing e_3' are p_9 and p_{12}. This way, e_3' is compared to each of its neighbors at least once as e_3' is grouped with its neighbors e_4' and e_5' in chunks $[p_9]$ and $[p_{12}]$ respectively.

Algorithm 1 describes the distribution of new entities over the chunks. It requires as input the list of new entities and the similarity threshold ϵ. The distribution of entities is performed in parallel and defines for each entity the chunks it is assigned to (line 1–5). The partial chunks created by each processing node are then merged to build the final chunks.

Entities of Δ_D^+ are distributed over chunks. As they can be in the neighborhood of entities of D, we will have to identify which entities of D have to be added to the generated chunks. This is the focus of the following subsection.

Algorithm 1. Distributing new entities

Input: the new dataset Δ_D^+, the similarity threshold ϵ
1: **for all** entity e' in Δ_D^+s **do in parallel**
2:   **for all** property $p \in cs(e')$ **do**
3:     $[p] = [p] \cup \{e'\}$
4:   **end for**
5: **end for**
6: Merge the chunks generated by the parallel execution for the same properties
Output: the chunks

Assigning Initial Entities to Chunks. As previously stated, the clusters that could be updated due to the insertion of new entities are those within the neighborhood of the new entities. Thus, the entities in D that are in the neighborhood of a newly inserted entity have to be identified. To this end, old entities that share common properties with the new ones are distributed over the generated chunks. By initial entities, we mean the entities in the dataset D prior to the addition of Δ_D^+, the set of new entities.

To distribute the entities in D, we first determine which properties have to be considered: for each entity $e \in D$, we compute its comparison set $cs(e)$ to select the properties to consider in order to determine which chunks it will be assigned to. The entities are assigned to the existing chunks according to their comparison set. Note that no new chunk is created and old entities are only assigned to the chunks created during the distribution of the new entities, according to the following rule:

$$if \; p \in cs(e) \; and \; \exists e' \in \Delta_D^+, e' \in [p] \; then \; e \in [p].$$

Example 2. In our running example, the result of the assignment of old entities, represented as green circles, is shown in Fig. 4.

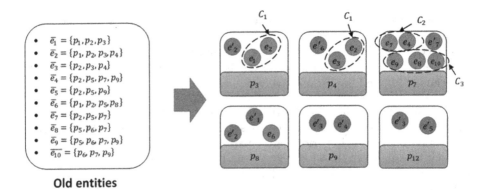

- $\bar{e}_1 = \{p_1, p_2, p_3\}$
- $\bar{e}_2 = \{p_1, p_2, p_3, p_4\}$
- $\bar{e}_3 = \{p_2, p_3, p_4\}$
- $\bar{e}_4 = \{p_2, p_5, p_7, p_9\}$
- $\bar{e}_5 = \{p_2, p_5, p_9\}$
- $\bar{e}_6 = \{p_1, p_2, p_5, p_8\}$
- $\bar{e}_7 = \{p_2, p_5, p_7\}$
- $\bar{e}_8 = \{p_5, p_6, p_7\}$
- $\bar{e}_9 = \{p_5, p_6, p_7, p_9\}$
- $\overline{e_{10}} = \{p_6, p_7, p_9\}$

Old entities

Fig. 4. The distribution of old entities

For example, the old entity e_4 described by $\overline{e_4} = \{p_2, p_5, p_7, p_9\}$ is assigned to the chunk $[p_7]$. Indeed, $cs(e_4) = \{p_5, p_7\}$, but the chunk $[p_5]$ has not been created while distributing the new entities, and therefore e_4 is only assigned to $[p_7]$. The resulting chunks after this assignment represent the final chunks.

Algorithm 2 presents the assignment of old entities to the chunks created by the distribution of entities within Δ_D^+. It requires the set of old entities D, the similarity threshold ϵ and the chunks already created.

The algorithm identifies the chunks to which each old entity should be assigned, according to both the properties describing the entities and the dissimilarity threshold (line 2). An entity e having a property p is assigned to the chunk $[p]$ if this chunk exists (line 3). Finally, the chunks generated in parallel by Algorithms 1 and 2 are merged to provide the final chunks.

Algorithm 2. Distributing entities

Input: the old entities D, the similarity threshold ϵ, the list of created chunks CH
1: **for all** entity $e \in D$ **do in parallel**
2:     **for all** property $p \in cs(e)$ **do**
3:         **if** $[p] \in CH$ **then**
4:             $[p] = [p] \cup \{e\}$
5:         **end if**
6:     **end for**
7: **end for**
8: Merge the chunks generated by the parallel execution for the same properties
Output: the chunks

The distribution principle used in this paper ensures that each new entity is grouped with all its candidate neighbors in $D \cup \Delta_D^+$. New entities are compared to all their candidate similar entities in order to define their neighborhood, then the clusters that should be updated or created are identified.

3.2   Computing the Neighborhood of the New Entities

In order to propagate the insertion of new entities into the existing schema, we need to compute the neighborhood of the new entities considering both the newly added entities and the old ones which have been previously assigned to existing clusters. This section first describes neighborhood computation for each new entity, then presents the identification of core entities in order to build the clusters.

As the chunks contain entities which are likely to be similar, the ϵ-neighborhood of a new entity is identified by computing the similarity between this new entity and all the other ones in the same chunk. We evaluate the similarity between two entities e_i and e_j using the *Jaccard index*.

Definition 6. *The ϵ-neighborhood of an entity e' is the set of entities similar to e' with a threshold of ϵ: neighborhood$_\epsilon(e') = \{e \in D \cup \Delta_D^+ \mid J(e', e) \geq \epsilon\}$*

According to the ϵ-neighborhood of the entities, we distinguish between three kinds of entities: *core entities* with at least *minPts* entities in their ϵ-*neighborhood*, *border entities*, that are not core entities but have at least one core entity in their ϵ-*neighborhood*, and *noise entities*, that are not core entities and have no core entity in their ϵ-*neighborhood*. These latters are not assigned to a cluster.

Definition 7. *An entity e' is a* core *entity if the number of entities within its ϵ-neighborhood is greater than the density threshold minPts, i.e.* $|neighborhood_\epsilon(e')| \geq minPts$.

The ϵ-neighborhood is computed for each new entity e' in each chunk by comparing e' to all the entities (new or old) within the same chunk. The ϵ-neighborhood is calculated in parallel within the different chunks, independently. Among the old entities, the only ones for which the ϵ-*neighborhood* is updated are the ones that are similar to a new entity. After updating the neighborhoods, an old entity that was a border may become a core entity, and an old entity that was a noise entity may become a border or a core entity.

Since the neighborhood of entities can be distributed over different chunks, the neighbors discovered in each chunk are consolidated and the list of neighbors for each entity in the whole dataset is built. This leads to the identification of the neighborhood of new entities within the whole dataset.

This process leads to the identification of the *core entities*, from which the clusters will be initiated; the cores are the entities having a number of neighbors greater or equal to *minPts*. The old border and noise entities that are similar to new ones can become core or border entities; adding new entities to their ϵ-*neighborhood* could make the number of their neighbors higher or equal to *minPts* and they will therefore become core entities, or they can be neighbors of a new core. As a consequence to such change occurring for an old entity, the clusters existing prior to the insertion of the new entities have to be updated.

The old entities that are not similar to a new one within a chunk are removed since they will not induce any change on the existing clusters: their initial clusters will not be updated, and they will not be assigned to any new cluster.

Example 3. In our example, ϵ is set to 0.7 and *minPts* is set to 2. The core entities are therefore e_2, e_4, e_9 within the old entities; and e'_1, e'_3, e'_7 within the new ones. The cores are represented by red borders in the Fig. 5.

For example, the entity e'_3 has two neighbors e'_4 and e'_5 respectively within the chunks $[p_9]$ and $[p_{12}]$.

The chunk $[p_3]$ is deleted since new entities within this chunk, such as e'_2, have no neighbors.

Algorithm 3 describes the computation of the neighborhood of new entities, executed in parallel withing each chunk.

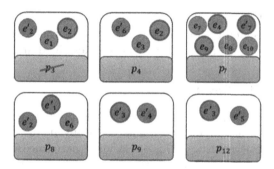

Fig. 5. Core identification within each chunk

Algorithm 3. Neighbors Computation

Input: the chunks CH, the similarity threshold ϵ, the density threshold $minPts$

1: **for all** $[p] \in CH$ **do in parallel**
2:   **for all** $e' \in [p] \mid e' \in \Delta_D^+$ **do**
3:     $neighborhood_\epsilon(e') = \{e \in [p] \mid J(e, e') \geq \epsilon\}$
4:   **end for**
5:   **for all** $e \in [p] \mid e \in D$ **do**
6:     $neighborhood_\epsilon(e) = neighborhood_\epsilon(e) \cup \{e' \in \Delta_D^+ \mid e \in neighborhood_\epsilon(e')\}$
7:   **end for**
8: **end for**
9: Merge the local neighborhoods to compute the complete list of neighbors of each entity
10: **for all** $e \in D$ **do**
11:   **if** $neighborhood_\epsilon(e) \cap \Delta_D^+ = \emptyset$ **then**
12:     $D = D \setminus \{e\}$
13:   **end if**
14: **end for**
15: **for all** $e \in D \cup \Delta_D^+$ **do**
16:   **if** $|neighborhood_\epsilon(e)| \geq minPts$ **then**
17:     $cores = cores \cup \{e\}$
18:   **end if**
19: **end for**
Output: cores

The algorithm first computes in parallel in each chunk the neighborhood of each new entity (line 2–4). If a new entity is similar to an old one, the neighborhood of the old entity is updated (line 5–7). Then, the lists of neighbors provided in each chunk are merged; the different calculating nodes exchange the partial lists of neighbors in order to build for each entity the list of its neighbors over the whole dataset. Old entities which do not have a new entity in their neighborhood are removed from the chunks (line 10–14). Finally, entities having a number of neighbors equal or greater than $minPts$ are identified as core entities (line 15–19).

By computing the neighborhood of new entities, we have identified the cores to build new clusters, and the old clusters impacted by the insertion of the entities. This restructuring of the existing clusters is described in the next section.

3.3   Generating the New Schema

To generate the new schema describing the whole dataset including both old and new entities, we first update the clusters locally in the chunks based on the neighborhood of new entities. This operation is performed in parallel within each chunk and it generates local clusters that contain similar entities within a single chunk, but it does not build clusters that span accross several chunks. Local clusters are therefore processed in order to detect the ones that have to be merged. Finally, the new schema is generated by propagating the updates applied on the old clusters over the entire dataset. We describe in this section the updates applied on the clusters within the chunks and the construction of the new schema.

Updating Clusters in Each Chunk. After adding a set of entities to a dataset D, the clusters discovered for D have to be updated to keep them coherent with new entities: (i) existing clusters could be updated by adding new elements, (ii) some clusters could be merged and (iii) new clusters could be created from new core entities.

In a density-based clustering algorithm, the clusters are built according to the density-reachability principle, introduced by the DBSCAN algorithm [11].

Definition 8. *An entity e is* directly density-reachable *from an entity e' wrt. ϵ and minPts if and only if e' is a core entity and e is in its ϵ-neighborhood, i.e. $|neighborhood_\epsilon(e')| \geq minPts$ and $e \in neighborhood_\epsilon(e')$.*

Definition 9. *An entity e is* density-reachable *from an entity e' wrt. ϵ and minPts if there is a chain of entities $e_1, \ldots, e_z$, $e_1 = e'$, $e_z = e$ such that e_{i+1} is directly density-reachable from $e_i, \forall i \in \{1, \ldots, z\text{-}1\}$.*

The elements to consider in our incremental algorithm are the new entities and the existing clusters within the neighborhood of new entities. It is proven that the incremental algorithm leads to the same result as the original DBSCAN algorithm [10].

By considering the new entities and their neighborhood, new density connections may be established. Based on the core entities within the chunks, computed during the previous stage, we can distinguish the following change operations on the existing clusters:

- If the ϵ-*neighborhood* of a new core $e' \in \Delta_D^+$ contains an old core entity $e \in D$ which belongs to an old cluster C, then the entity e' is assigned to C and C is also expanded with entities that are density-reachable from e'.

- If a core entity $e \in D \cup \Delta_D^+$ has no old core entity in its ϵ-neighborhood, then a new cluster is created and the entities that are density-reachable from e are added to this cluster.
- If the ϵ-*neighborhood* of a core entity $e \in D \cup \Delta_D^+$ contains two or more old core entities, which belong to distinct clusters, then these clusters are merged and the resulting cluster is expanded with the entities that are density-reachable from e.
- If an old core entity has a new entity which is not a core within its neighborhood, then the corresponding new entity is absorbed by the cluster containing this old core entity.

Note that the number of cores is lower than the total number of entities within a chunk. Therefore, iterating over the cores instead of all the entities improves the efficiency of the process.

During this stage, we update the clusters having an entity in the neighborhood of a newly inserted entity. This is done according to the rules defined above. These rules are executed in parallel in the different chunks based on the neighborhood of the entities previously computed. To prevent ambiguity between clusters of different chunks, the clusters are denoted by the ids of the chunks followed by an index.

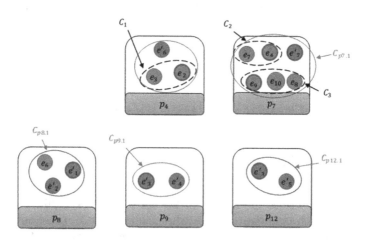

Fig. 6. Clustering the chunks

Example 4. Updating the clusters within the chunks produces the result presented in Fig. 6. For example, the entity e_6' is absorbed by the existing cluster C_1. The cluster $C_{p8.1}$ is created from the new core entity e_1' and its neighbors. The entity e_7' is a core and its neighborhood contains two old core entities e_4 and e_9. The clusters C_2 and C_3 that contain respectively e_4 and e_9 are merged into a new cluster $C_{p7.1}$. The core entity e_3' creates two different clusters $C_{p9.1}$

and $C_{p_{12.1}}$ within the chunks $[p_9]$ and $[p_{12}]$ respectively. The clusters that span over different chunks such as $C_{p_{9.1}}$ and $C_{p_{12.1}}$ will be merged during the next stage.

Algorithm 4. New Local Clusters

Input: CH: the chunks, $Cores$: the core entities having a new entity in their neighborhood

1: **for all** $[p] \in CH$ **do in parallel**
2:     is-visited $= \emptyset$
3:     **for all** $e \in Cores$ **do**
4:         **if** $e \notin$ is-visited **then**
5:             is-visited $=$ isVisited $\cup \{e\}$
6:             **if** $e.cluster \neq null$ **then**
7:                 $C = e.cluster$
8:             **else**
9:                 Create a new cluster $C = \{e\}$
10:            **end if**
11:            $C = C \cup neighborhood_\epsilon(e)$
12:            **for all** $e' \in C \mid e' \in cores$ and $e' \notin$ is-visited **do**
13:                **if** $e'.cluster = null$ **then**
14:                    $C = C \cup \{e'\} \cup neighborhood_\epsilon(e')$
15:                **else**
16:                    $C' = e'.cluster$
17:                    $C = C \cup C'$
18:                **end if**
19:            **end for**
20:        **end if**
21:        local-clusters $=$ local-clusters $\cup\, C$
22:    **end for**
23: **end for**
Output: local-clusters

Algorithm 4 describes the update of the set of clusters within each chunk. It iterates over each core entity within the chunks (line 3); these core entities could be new entities or old ones that have a newly inserted entity in their neighborhood. Then, it either identifies the cluster of the current core in order to expand it (line 6–7), or create a new cluster for this core (line 8–9), and the cluster is expanded by adding its neighbors (line 11). Next, the algorithm identifies among the added neighbors those which are core (line 12), and adds their neighbors to the cluster if these latter do not belong to another cluster (line 13–14). If the created cluster C contains a core entity that belongs to another old cluster C', then these clusters are merged (line 16–17).

At the end of this stage, clusters are produced in each chunk. However, these clusters are not the final result. Indeed, clusters can span across several chunks since they are built independently. In addition, the clusters generated from the new core entities and their neighborhood have to be propagated into the existing schema. In the next stage, we will describe the process of building the final clustering result from the local clusters in each chunk. This stage is processed in a single node and provides the final clustering result.

Generating the Final Clusters. Our distribution principle may result in clusters that span across different chunks. These clusters are the ones that share a core entity, and they will be merged into a single cluster. Besides, the entities clustered in a given chunk are either new entities, or old entities that are in the neighborhood of new ones. In order to provide the final result, the updates applied on the clusters have to be propagated on the old entities which have not been distributed on the chunks and not considered during the clustering.

In this section, we first describe the identification of the clusters that span across several chunks, and the way the corresponding local clusters are merged. This process is executed on one computing node and is not parallelized. Then, we present the generation of the new schema according to the generated clusters.

According to the density-based clustering algorithm, an entity e is assigned to a cluster C if e is density-reachable from a core entity in C. If this same entity e is also in another local cluster C', e is also density-reachable from a core entity in C'. If e is a core, it represents a bridge between the entities in the clusters C and C', making them density-reachable. The clusters that span across several chunks are identified by finding out the local clusters, produced within each chunks, that share a common core entity within the newly inserted entities. These clusters are merged to produce the final result.

If a border entity is assigned to different clusters during the clustering, it would be randomly assigned to one of these clusters during this stage.

After producing the final clusters representing the new entities and their neighborhood, this result is propagated in the old clusters to construct the new schema. The old clusters to consider in this stage are those which have been merged as a consequence to the insertion of a new core in their neighborhood. The entities previously assigned to these old clusters should therefore be re-assigned to the new cluster resulting from their merge.

If two old clusters C_i and C_j are merged to produce a new cluster C', all the elements of these clusters should be assigned to C'. However, during the distribution of old entities, not all the old entities are distributed on the chunks. We therefore need to change the assignment of old entities which have not been distributed in the chunks and which belong to clusters that have been merged into a new cluster.

Example 5. In our example, the clusters C_2 and C_3 presented in Fig. 7 are merged since they are in the neighborhood of the core entity e_7' and the id of the created cluster is C_1'. This implies that all the entities in these clusters, such as the entity e_7, will change their cluster assignment to C_1'.

Finally, all the entities that are not assigned to a cluster, are considered as noise.

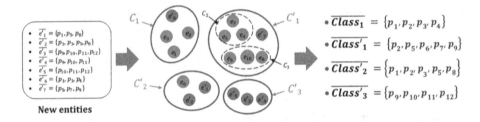

$\overline{Class_1} = \{p_1, p_2, p_3, p_4\}$

$\overline{Class'_1} = \{p_2, p_5, p_6, p_7, p_9\}$

$\overline{Class'_2} = \{p_1, p_2, p_3, p_5, p_8\}$

$\overline{Class'_3} = \{p_9, p_{10}, p_{11}, p_{12}\}$

Fig. 7. Updating the schema after the insertion of new entities

This stage provides the final clustering result, ensuring that the clusters are the same as the ones that a sequential DBSCAN algorithm executed on the whole dataset would produce.

Example 6. Figure 7 presents the updates on the classes introduced in Fig. 1 following the insertion of a set of new entities. For instance, the class $Class_1$ also represents the new entity e'_6 within the corresponding cluster C_1. The classes $Class_2$ and $Class_3$ are merged into $Class'_2$ since the corresponding clusters C_2 and C_3 have a common core e'_7 that is similar to one of their entities, e_4 and e_9 respectively. Additionally, new classes ($Class'_2$ and $Class'_3$) are created, representing the newly generated clusters.

Algorithm 5 describes the construction of the final clustering result. First, it iterates over the new core entities. If a core entity belongs to two distinct clusters or more, then these clusters are merged (line 2–5). During this iteration, the old clusters which have changed their id are identified and saved (line 6–10). If two clusters have been merged, the cluster id of all the entities belonging to these clusters is changed into the one of the cluster resulting from the merge (12–16).

4   Schema Evolution After Entity Deletions

In this section, we introduce an approach that allows to keep the schema coherent with the dataset when entities are deleted. In order to efficiently manage incrementally evolving big datasets, the clustering is restricted to deleted entities and their neighborhood. Updating the clusters containing deleted entities ensures providing the same result as executing DBSCAN on the global dataset [10].

Our approach is composed of three main steps, as illustrated in Fig. 8, parallelized and implemented using a big data technology.

First, the data are split into partitions and distributed overs different processes in order to effectively manage big datasets. The partitions contain the entities within the neighborhoods of deleted entities and the entities impacted by the modifications. The impacted entities and the deleted entities are distributed through distinct partitions according to their cluster, thus, all the modifications related to a cluster are applied within the same partition.

Algorithm 5. Restructuring the Classes

Input: *local-clusters*: the local clusters, *Cores*: the core entities, *oldClusters*: the old
   entities not in the chunks and their clusters

1: *clusters = local-clusters*
2: *newIds* : $HashMap[oldId, newId] = \emptyset$
3: **for all** $e \in D \mid e \in Cores$ **do**
4:    $lc_e = \{C \in clusters \mid e \in C\}$
5:    $clusters = clusters \setminus lc_e \cup (\cup_{C \in lc_e} C)$
6:    **for all** $c_i \in lc_e$ **do**
7:       **if** $c_i \in oldClusters$ **then**
8:          $newIds = newIds \cup (c_i, C)$
9:       **end if**
10:    **end for**
11: **end for**
12: **for all** $e \in oldClusters$ **do**
13:    **if** $e.cluster \in newIds$ **then**
14:       $e.cluster = newIds.get(e.cluster)$
15:    **end if**
16: **end for**

Output: clusters

After the distribution, the neighborhood of the distributed entities are updated by removing the deleted entities. The number of neighbors for each entity is then evaluated in order to define the status of each impacted entity (core, border or noise).

Finally, based on the neighborhood of the impacted entities, the existing clusters are updated in each partition: (i) a cluster could be deleted if all its core entities are deleted, (ii) a cluster could loose some of its entities, (iii) a cluster could be split into distinct clusters. The clusters updated in each partition are then integrated to the initial clusters to generate the new clusters which represent the classes of the new schema.

We detail in the following sections the different steps of our proposal.

4.1   Data Distribution

The update of a schema describing an RDF dataset requires the identification of the clusters impacted by the deletion of entities. These clusters are those having elements which are neighbors of a deleted entity. In a set of entities D, an entity e is impacted by the deletion of an entity e' if e is within the neighborhood of e'.

The changes of the clusters are therefore restricted to the neighborhood of the deleted entities because the status of these entities can change. As core entities may become non-core, density connections may be lost which implies the modification of the clusters. The status of an entity which is not within the neighborhood of a deleted entity does not change and does not imply any modification on its cluster. Updating the clusters within the neighborhoods of the deleted entities ensures providing the same result as executing DBSCAN on

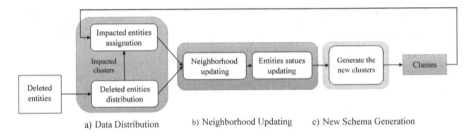

a) Data Distribution   b) Neighborhood Updating   c) New Schema Generation

Fig. 8. Overview of our incremental schema discovery approach for dealing with deletions

the global data. In our approach, we consider only the impacted entities in order to propagate the updates in the schema [10].

In order to process big datasets, we distribute the impacted entities into different partitions and process the modifications on different computing nodes. The modification of the clusters is then performed in parallel within each partition by considering only the impacted entities.

A partition of data is defined as follows:

Definition 10. *A partition denoted by d_i is a subset of entities that belong to the cluster C_i and which are within the neighborhood of a deleted entity:*

$$d_i = \{e | e \in C_i, neighborhood_\epsilon(e) \cap \Delta_D^- \neq \emptyset\}$$

Since the neighborhood of the entities is known, our approach groups each deleted entity with its neighbors in a same partition. Then, the neighborhoods of impacted entities are updated by removing the deleted ones. The distribution of entities in order to update the clusters after a deletion of a set of entities aims to create different processes that would be executed in parallel.

In our incremental approach, we first distribute the set of deleted entities Δ_D^- through different partitions according to their clusters. Initially, each partition contains a set of deleted entities which belong to the same cluster. Then, the set of entities D is distributed through the created partitions based on the neighborhood of the entities. An entity e is assigned to a partition d_i which contains a deleted entity within the neighborhood of e: e is assigned to a partition d_i if $e' \in d_i, e' \in \Delta_D^-$ and $e \in neighborhood_\epsilon(e')$.

The deleted entities which were not previously assigned to a cluster are not distributed in the partitions since their deletion does not affect any existing cluster.

Example 7. In our example, we consider the dataset $D \cup \Delta_D^+$ presented in Fig. 7 and the corresponding schema. We assume that the set of entities $\Delta_D^- = \{e_1', e_2', e_3', e_4', e_5', e_6', e_7'\}$ is deleted from D.

The distribution of the set of deleted entities Δ_D^- produces the partitions presented in Fig. 9. For example, the deleted entities e_1' and e_2' which belong to the cluster C_2' are assigned to the partition $d_{C_2'}$.

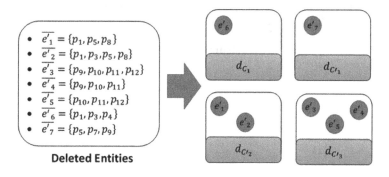

$$\bullet \ \overline{e'}_1 = \{p_1, p_5, p_8\}$$
$$\bullet \ \overline{e'}_2 = \{p_1, p_3, p_5, p_8\}$$
$$\bullet \ \overline{e'}_3 = \{p_9, p_{10}, p_{11}, p_{12}\}$$
$$\bullet \ \overline{e'}_4 = \{p_9, p_{10}, p_{11}\}$$
$$\bullet \ \overline{e'}_5 = \{p_{10}, p_{11}, p_{12}\}$$
$$\bullet \ \overline{e'}_6 = \{p_1, p_3, p_4\}$$
$$\bullet \ \overline{e'}_7 = \{p_5, p_7, p_9\}$$

Deleted Entities

Fig. 9. Distribution of the deleted entities into partitions

Then, the entities impacted by the deletion are assigned to the created partitions. The result of the assignment of old entities, represented as green circles, is shown in Fig. 10. For example, the entity e_2 is assigned to the partition d_{C_1} since it is within the neighborhood of the entity e'_6.

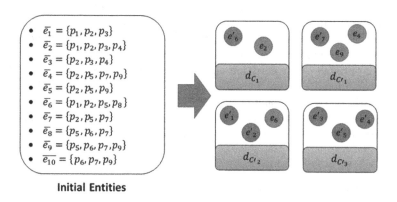

$$\bullet \ \overline{e}_1 = \{p_1, p_2, p_3\}$$
$$\bullet \ \overline{e}_2 = \{p_1, p_2, p_3, p_4\}$$
$$\bullet \ \overline{e}_3 = \{p_2, p_3, p_4\}$$
$$\bullet \ \overline{e}_4 = \{p_2, p_5, p_7, p_9\}$$
$$\bullet \ \overline{e}_5 = \{p_2, p_5, p_9\}$$
$$\bullet \ \overline{e}_6 = \{p_1, p_2, p_5, p_8\}$$
$$\bullet \ \overline{e}_7 = \{p_2, p_5, p_7\}$$
$$\bullet \ \overline{e}_8 = \{p_5, p_6, p_7\}$$
$$\bullet \ \overline{e}_9 = \{p_5, p_6, p_7, p_9\}$$
$$\bullet \ \overline{e}_{10} = \{p_6, p_7, p_9\}$$

Initial Entities

Fig. 10. Distribution of impacted old entities

Algorithm 6 describes the distribution of the entities over the partitions. The first loop (line 1–4) distributes the deleted entities, it defines for each entity the partition it is assigned to based on its cluster. The second one (line 5–9) distributes the entities within the neighborhood of a deleted entity. The distribution is performed in parallel and produces partial partitions which are then merged to build the final partitions.

The distribution of the entities groups in the same partition all the deleted entities and their neighborhood. It allows to update in parallel, within each partition the neighborhoods of the impacted entities, and to update the clusters.

Algorithm 6. Distributing the deleted entities

Input: the deleted entities Δ_D^-, the dataset D
1: **for all** entity e' in Δ_D^- **do in parallel**
2:    $i = e'.getCluster()$
3:    $d_i = d_i \cup \{e'\}$
4: **end for**
5: **for all** entity e in D **do in parallel**
6:    **if** $neighborhood_\epsilon(e) \cap \Delta_D^- \neq \emptyset$ **then**
7:       $i = e.getCluster()$
8:       $d_i = d_i \cup \{e\}$
9:    **end if**
10: **end for**
11: Merge the partitions generated by the parallel execution for the same cluster
Output: the partitions

4.2 Updating the Neighborhood of the Impacted Entities

In order to update the schema after the deletion of entities, the neighborhood of the entities impacted by the modifications has to be updated. The impacted entities are those within the neighborhood of the deleted ones.

According to our assumptions, the neighborhood of the entities is known and its computation is not required. When entities are deleted, they are only removed from the neighborhood of the impacted entities.

In our approach, each partition contains all the entities needed in order to update a considered cluster. This distribution allows a parallel modification of the neighborhood of the impacted entities. The list of neighbors of each impacted entity is updated by removing the deleted entities within each partition. For each entity e, we delete the entities within Δ_D^- from its list of neighbors.

$$\forall e \in D, neighborhood_\epsilon(e) = neighborhood_\epsilon(e) \setminus \Delta_D^-.$$

When deleting a set of entities Δ_D^-, impacted entities can change their status in two different cases:

- An entity e becomes a border entity if the number of entities within its neighborhood after the deletion of Δ_D^- is lower than the $minPts$ parameter, and if e is within the neighborhood of a core.
- An entity e becomes a noise entity if it is not a core and it is not within the neighborhood of a core after the deletion of Δ_D^-.

According to the list of neighbors for each entity, we identify the core entities which become border entities and the clusters which should be updated. Deleting entities from the neighborhood of an entity could change its status. The old border entities could loose the cores within their neighborhood and become noise entities. Deleting entities from the neighborhood of a core entity e could make the number of its neighbors lower than the density threshold $minPts$ and therefore e becomes either a border or a noise entity.

As a consequence to such changes occurring for an impacted entity, the clusters existing before the deletion have to be updated.

The old entities that are not within the neighborhood of a deleted entity are not considered since their status remain the same and they will not induce any change on the existing clusters.

Example 8. The border entity e_6 is updated by deleting the entities e'_1 and e'_2: $neighborhood_\epsilon(e_6) = neighborhood_\epsilon(e_6) \setminus \{e'_1, e'_2\}$. The entity e_6 becomes a noise entity since there is no other core entity in its neighborhood.

Algorithm 7 describes the update of the neighborhood of each entity within each partition. It iterates over the entities within the partitions (line 2), then the deleted entities are removed from the neighbors list of each entity (line 3).

Algorithm 7. Neighborhood updating

Input: D: the partitions, Δ_D^-: the deleted entities
1: **for all** $d_i \in D$ **do in parallel**
2:    **for all** $e \in d_i$ **do**
3:        $neighborhood_\epsilon(e) = neighborhood_\epsilon(e) \setminus \Delta_D^-$
4:    **end for**
5: **end for**

At the end of this stage, the neighborhood of each impacted entity is updated, which allows to update the clusters.

4.3 Generating the New Schema

To generate the new schema describing the dataset after deleting a set of entities, the clusters are updated based on the neighborhood of the impacted entities. The updates are performed in parallel within each partition and produce the new clusters that should be integrated with old clusters in order to build the new classes of the schema.

After deleting a set of entities, density connections may be removed, and a chain of density-reachable entities could be broken. This will require the modification of the clusters containing these entities.

When deleting a set of entities Δ_D^- from a dataset D, the clusters representing D have to be updated to keep them coherent: (i) existing clusters could loose some entities, (ii) clusters could be split into different clusters and (iii) some clusters could be deleted. Based on the updates occurring on the neighborhood of the entities within the partitions, we can distinguish the following change operations on the existing clusters:

– If a core entity e within a cluster C becomes non-core, and the core entities within the neighborhood of e are density-reachable from each other, then, the cluster C is reduced by removing the border entities within the neighborhood of e which are not density-reachable from another core entity within C.

- If a core entity e within a cluster C becomes non-core, and the core entities within its neighborhood are no longer density-reachable from each other, then C is split to generate several clusters based on the density-reachability of its entities.
- If a border entity e within a cluster C is deleted, the cluster is reduced by removing e. The density of the core entities in the neighborhood of e are checked in order to identify the ones that are no longer core entities.

During this stage, the clusters in the neighborhood of the deleted entities are updated according to the rules defined above. These rules are executed in parallel in the different partitions based on the neighborhood of the entities which have been updated as described in Sect. 4.2.

For each existing core entity e in a cluster C which is either deleted or no longer a core, the density connections within C have to be checked to determine whether all the entities are density-reachable from each other.

If two core entities in the neighborhood of e are no longer density-reachable from each other, C is split into different clusters $\{C_1, \cdots, C_n\}, n \geq 2$, according to the principles of DBSCAN. As a consequence, the entities which were assigned to C are assigned to one of the generated clusters C_i. In order to assign the entities to the new clusters, the core entities within C are uploaded and assigned to a cluster C_i based on the density-reachability principle. A core e_i is assigned to a cluster C_i if C_i contains a core entity e_j such that $e_i \in neighborhood_\epsilon(e_j)$. Then, the neighbors of each core are added to the cluster.

If the cores within the neighborhood of e are density-reachable from each other, then each border entity e_k in the neighborhood of e is checked to determine if it is density-reachable from another core entity in C, which is the case if there is another core entity in the neighborhood of e_k. Otherwise, the border entity e_k becomes a noise entity.

If a border entity e_k is deleted or becomes a noise entity, the number of neighbors of the core entities within the neighborhood of e_k is computed in order to identify the core entities that become non-core.

Finally, if a cluster C has no core entities, either because they have been deleted or they have become non-core, then the cluster C is deleted.

The final set of clusters contains both the restructured clusters and the old clusters which have not been modified. These clusters represent the classes of the schema.

Example 9. Considering the set of initial clusters provided in Fig. 7, Fig. 11 represents the clusters generated after the deletion of the set of entities $\Delta_D^- = \{e_1', e_2', e_3', e_4', e_5', e_6', e_7'\}$.

For example, the cluster C_1' is split into C_1'' and C_2'' since the core entity e_7' is deleted. The clusters C_1'' and C_2'' are built by assigning the entities within C_1' to one of the clusters based on the density principle. The cluster C_2' is deleted since all its cores have been deleted, and the entity e_6 becomes a noise entity.

Algorithm 8 describes how the clusters are updated in order to reflect the deletion of the entities. In parallel within the partitions and for each entity, the

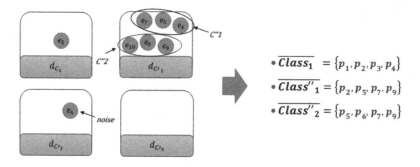

$$\bullet \overline{Class_1} = \{p_1, p_2, p_3, p_4\}$$

$$\bullet \overline{Class''_1} = \{p_2, p_5, p_7, p_9\}$$

$$\bullet \overline{Class''_2} = \{p_5, p_6, p_7, p_9\}$$

Fig. 11. Generating the new clusters

algorithm checks if a core entity becomes non-core (line 3). In that case, it checks if the cores are density-reachable from each other. It selects a random core e, and computes the number of cores, denoted nb, that are density-reachable from e (line 5–7). This number is compared to the number of cores within the partition (line 10), denoted by nc. If $nb < nc$, the cluster is split into different clusters according to the principles of DBSCAN and its entities are re-assigned to the new clusters (line 11–19). If the impacted entity is a border and there is no core within its neighborhood, it becomes a noise entity (line 21–25).

5   Experimental Evaluations

In this paper, our experiments are focused on the performances of our approach when applied to large evolving datasets. We evaluate the efficiency of our incremental density-based clustering algorithm compared to the scalable DBSCAN algorithm proposed in [5], when a dataset evolves by adding or deleting sets of entities. Both algorithms rely on the Apache Spark 2.0 framework. In our experiments, we present a comparison between these algorithms and derive the speed-up factor when using our incremental approach to reflect the insertion or deletion of sets of entities in the clustering result instead of using the scalable DBSCAN algorithm on the dataset composed of the old entities and the newly inserted ones in the case of insertions, and the old entities minus the deleted ones in the case of deletions. We have used our implementation of the scalable DBSCAN algorithm, available online[2].

As previously explained in this paper, clustering a dataset using our incremental approach provides the same result as executing the DBSCAN algorithm on the whore dataset. This feature of our approach is important since it preserves the good quality of the extracted schema. Previous works have shown that extracting a schema from an RDF dataset using DBSCAN provides a good quality result, with good precision and recall, and detects classes which were not declared in the dataset [5,24].

---

[2] https://github.com/BOUHAMOUM/SC-DBSCAN.

Algorithm 8. Clusters Updating

Input: D: the partitions, $Cores$: the old core entities, $minPts$: density threshold, $clusterID$: the impactedCluster

1: **for all** $d_i \in D$ **do in parallel**
2:    **for all** $e \in d_i$ **do**
3:      **if** $e \in Cores$ AND $neighborhood_\epsilon(e) < minPts$ **then**
4:        denseSet = {}
5:        **for all** $e' \in neighborhood_\epsilon(e)$ AND $e' \in Cores$ **do**
6:          $denseSet = denseSet \cup (neighborhood_\epsilon(e') \cap Cores)$
7:        **end for**
8:        $coreOfE = neighborhood_\epsilon(e) \cap Cores$
9:        **if** $|denseSet| \neq |coreOfE|$ **then**
10:         $impactedCluster = load(clusterID)$
11:         **for all** $e \in impactedCluster$ **do**
12:           **if** $e.isCore()$ **then**
13:             Create a new cluster $C = \{e\} \cup neighborhood_\epsilon(e)$
14:             **for all** $e' \in C \mid e' \in cores$ **do**
15:               $C = C \cup \{e'\} \cup neighborhood_\epsilon(e')$
16:               $e'.cluster = C$
17:             **end for**
18:           **end if**
19:         **end for**
20:       **else**
21:         **for all** $e' \in neighborhood_\epsilon(e)$ **do**
22:           **if** $neighborhood_\epsilon(e') \cap Cores = \emptyset$ **then**
23:             $e'.cluster = null$
24:           **end if**
25:         **end for**
26:       **end if**
27:     **end if**
28:   **end for**
29: **end for**

Each time a set of entities Δ_D^+ is added to the initial dataset D or a set of entities Δ_D^- is deleted, we evaluate the execution time needed by our incremental algorithm to update the clustering result obtained on D so as to reflect the insertion of Δ_D^+ or the deletion of Δ_D^-. The execution time of this scenario is compared to the execution time needed by the scalable DBSCAN algorithm in order to cluster the dataset composed of both the initial dataset and the inserted set of entities, i.e. $D \cup \Delta_D^+$, or the initial dataset minus the set of deleted entities, i.e. $D \setminus \Delta_D^-$.

In this section, we first present our experimental setup. Then, we evaluate our incremental approach for schema discovery applied on a dataset that evolves by inserting new entities. Finally, we evaluate our approach for schema updating when entities are deleted from the initial dataset.

5.1 Experimental Setup

In this subsection, we present the datasets used to evaluate both the scalability of our algorithms and their behaviour when dealing with insertions or deletions of sets of entities of various sizes. Then, we present the experimental environment.

In our experiments, we have first used a synthetic multidimensional dataset of 4 millions entities, generated using "IBM Quest Synthetic Data Generator" [18]. This well known generator was heavily used in the data mining community to evaluate the performances of frequent itemset mining algorithms. In our context, the generator produces the properties of each entity that will be used in our experiments.

Then, as the complexity of our incremental approach depends on the number of inserted or deleted entities, we have therefore evaluated the incremental algorithm by inserting or deleting sets of entities of different sizes.

Finally, we illustrate the efficiency of our approach on real datasets. To this end, we apply our approach on 1.2 million entities extracted from DBpedia[3] [2]. DBpedia is a project aiming to extract structured content from the information created in the Wikipedia project and to make it available on the Web. DBpedia allows users to query relationships and properties of Wikipedia resources, including links to other related datasets.

All the experiments have been conducted on a cluster running Ubuntu Linux consisting of 5 nodes (1 master and 4 slaves), each one equipped with 30 GB of RAM, a 12-core CPU.

5.2 Evaluating Scalability When Dealing with Insertions

We have first evaluated the scalability of our approach and compared it to the scalable DBSCAN algorithm using several synthetic datasets where we have added datasets of different sizes. Figures 12a, 12b and 12c show both algorithms' runtime as a function of the dataset size. The scalable DBSCAN takes as input the global dataset while the incremental algorithm takes as input the clusters of the previous execution and the newly inserted entities.

The results show that clustering a small dataset is faster using the scalable DBSCAN than using the incremental DBSCAN. This is due to the fact that clustering a small number of entities is very fast and requires a few seconds (22 s to clusters 200k entities). Besides, the incremental algorithm executes extra operations such as the assignment of old entities and the union of the result produced by this assignment with the chunks created during the distribution of the new entities, which makes it slower on small datasets compared to the scalable algorithm.

However, when the number of entities is higher, clustering a dataset using the incremental DBSCAN algorithm is faster. This is due to the fact that the clustering is applied on new entities and their neighborhoods, which counterbalances the extra operations, while the scalable DBSCAN algorithm has to build

---

[3] http://downloads.dbpedia.org/3.9/.

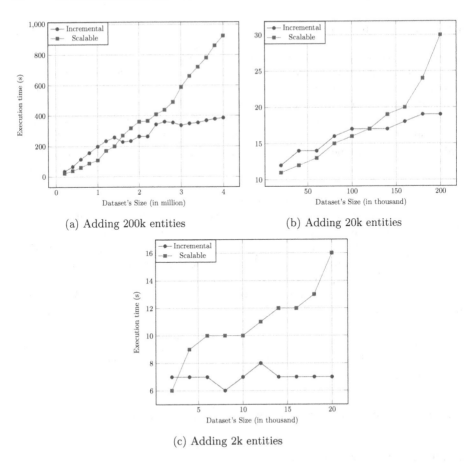

(a) Adding 200k entities

(b) Adding 20k entities

(c) Adding 2k entities

Fig. 12. Incremental vs. sequential scalable algorithm

the clusters by computing the neighborhood of all the entities, which is a more expensive operation. In addition, the incremental approach produces a lower number of new clusters compared to the scalable algorithm. Thus, when merging the clusters determined within each chunk, a process which is executed in one node, the incremental algorithm has to deal with a lower number of clusters which makes it faster.

We can observe that the bigger the dataset, the larger the gap between the execution time of both algorithms, and the higher the gain achieved by the incremental approach.

Since the complexity of the incremental algorithm is defined by the number of new entities and their neighborhood, we have experimented the insertion of sets of entities Δ_D^+ of different sizes. The results show that the benefits of the incremental algorithm compared to the scalable DBSCAN algorithm vary according to the size of the added set of entities. The smaller the sets of added entities, the faster the clustering using the incremental algorithm. In our experiments, when adding 200k entities at each step, the incremental algorithm becomes faster than the scalable algorithm when the whole dataset reaches the size of 1.6M entities, while when adding 20k at each step, it becomes faster when the dataset reaches the size of 140k entities (Fig. 12b). When the size of the inserted datasets is smaller, the gain achieved by the incremental algorithm is more important, as shown in Fig. 12c after the second insertion. These results are explained by the fact that the incremental algorithm generates the clusters only for the new entities and their neighborhood. It does not take into consideration all the dataset. The smaller the inserted set of entities, the fewer the number of entities which have to be managed by the algorithm, which makes its execution faster.

Finally, we have evaluated the efficiency of our approach on real datasets. Figure 13 illustrates the ability of our incremental algorithm to cluster real datasets, such as DBpedia, a large RDF source from which we have extracted more than 1.2 million patterns. The patterns represent all the existing combinations of properties describing the entities in the dataset. Each pattern is a combination of properties for which there is at least one instance in the dataset. Entities having exactly the same property sets are represented by a single pattern. To extract the patterns, we have used the approach proposed in [7]. Considering patterns instead of entities reduces the size of the input data and helps speeding up the clustering. Similar to the evaluations on the synthetic datasets, we have added in each insertion to the initial dataset D, a set of entities Δ_D^+ containing 100k entities. Then the execution time of the incremental algorithm is compared to the scalable DBSCAN when executed on the entire dataset.

This evaluation shows that the incremental algorithm overcomes the scalable algorithm in terms of performances. In addition, entities in DBpedia have a high number of properties; some entities have more than 600 properties. As a consequence, the scalable algorithm creates big sized chunks; this has a negative impact on its performances because it reaches the calculation's limit of the cluster when computing the ϵ-neighborhood of the entities, as we notice on the dataset having 1 million entities. However, the incremental algorithm is not impacted by entities having a high number of properties since it manages in each clustering only a limited subset of the dataset and computes the ϵ-neighborhood for the new entities only.

5.3   Evaluating Scalability When Dealing with Deletions

Similarly to the experiment showing the efficiency of our approach to deal with insertions, we have evaluated the scalability of our approach for dealing with entity deletions, and compared it to the scalable DBSCAN using synthetic datasets of different sizes. Figures 14a, 14b and 14c show both algorithms' runtime as a function of the dataset size. The scalable algorithm takes as input the global dataset while the incremental algorithm takes as input the clusters of the previous execution and the set of deleted entities.

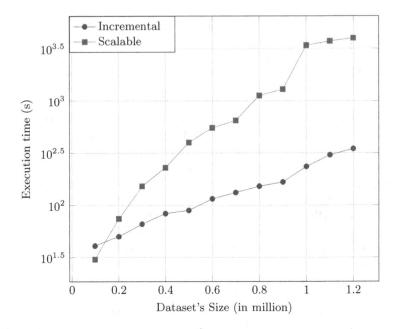

Fig. 13. Clustering DBpedia (managing insertions)

The results show that the incremental algorithm overcomes the scalable algorithm in terms of performances when dealing with the deletion of sets of entities. The scalable algorithm requires the computation of the neighborhood of each entity, which represents a complex operation. However, as the neighborhoods of entities are known when executing the incremental algorithm, they are only updated by considering the deleted entities. As a consequence, the incremental algorithm does not need the computation of the similarity between entities and skips the costly operations performed by the scalable algorithm. In addition, the incremental algorithm manages a subset of the initial dataset for each deletion, and it is not impacted by the size of the initial dataset. Furthermore, each partition contains all the required information to update a cluster. As consequence, our incremental approach does not require any communications between the computing nodes and thus does not apply any shuffle operations in order to compute the clusters.

We have also experimented the deletion of sets of entities Δ_D^- of different sizes. The results show that the smaller the sets of deleted entities, the faster the clustering using the incremental algorithm. Indeed, the incremental algorithm manages only the entities impacted by the deleted ones. Thus, when deleting a small set of entities, the incremental algorithm processes a smaller number of entities which makes its execution fast.

Finally, we have evaluated the efficiency of our approach on real datasets. Figure 15 represents the evaluations of both incremental and scalable algorithms

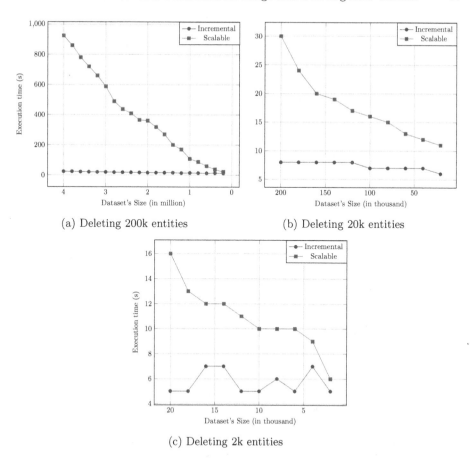

(a) Deleting 200k entities

(b) Deleting 20k entities

(c) Deleting 2k entities

Fig. 14. Incremental vs. sequential scalable algorithm when dealing with deletions

used to compute the clusters on different subsets of DBpedia. For the incremental algorithm, a set Δ_D^- containing 100k patterns is deleted for each evaluation.

This evaluation shows that the incremental algorithm overcomes the scalable one when applied on real datasets to reflect the updates on the schema. Similarly to the evaluations on synthetic datasets, the incremental algorithm processes in each deletion the entities within the neighborhood of deleted ones, compared to the scalable algorithm which considers the whole dataset. In addition, the incremental algorithm does not compute the similarity between the entities which makes it very fast.

As a conclusion to the experiments, we have shown that the incremental algorithm overcomes the scalable one both for dealing with insertions and for dealing with deletion. Indeed, the incremental algorithm updates the set of clusters by considering the smallest subset of the initial dataset consisting only on the impacted entities. Thus, the incremental algorithm process the data faster and presents better performances.

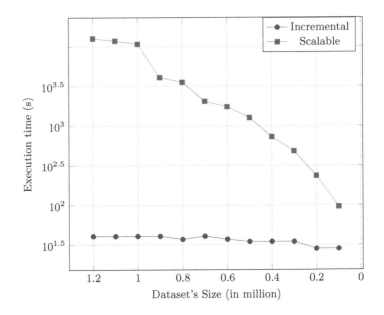

Fig. 15. Clustering DBpedia (managing deletions)

6   Related Work

Several approaches have been proposed for schema discovery in RDF datasets. A comprehensive survey of existing approaches is provided in [22]. Some of these approaches have used clustering algorithms to group similar entities in order to form the classes representing the schema. Among these works, the approaches presented in [23,24] have used density-based clustering algorithms and have adapted them to generate classes and links between them. The approach described in [8] relies on hierarchical clustering for generating the underlying types in an RDF dataset. The work presented in [33] uses the FP-Growth algorithm to find the most frequent properties describing a schema based on the classes chosen by the user. However, these approaches have not dealt with scalability issues, and most of them do not scale up to process very large datasets.

To manage the incrementality issues, the approach presented in [24] proposes a supervised learning step in order to define the type of a new incoming entity, by introducing the concept of fictive entity representing a class, and by comparing a new entity to each fictive entity to determine its type. However, the goal of this approach is to assign an existing type to an instance, and it does not generate new types. HInT, the approach presented in [21] proposes an incremental type discovery for RDF data sources based on locality sensitive hashing algorithm [20]. The approach incrementally indexes the entities and groups them to identify the types within an RDF dataset. It is natively scalable. However, it does not deal with the deletion of existing entities.

Some approaches have specifically addressed the scalability of schema discovery, providing algorithms able to manage large datasets implemented using a big data technology such as Hadoop [12] or Spark [34]. However, unlike our approach, these algorithms rely on type declarations to group entities into classes, and then provide a representative schema to help understand the data [3, 4, 30]. Such approaches can not be used when these declarations are not provided in the dataset.

Our clustering algorithm is inspired by DBSCAN, which is well suited to the requirements of RDF datasets. This is mainly because it produces clusters of arbitrary shape, which is important in our context where entities of the same type can be described by heterogeneous property sets. Furthermore, it does not require as an input the number of resulting clusters, and it detects noise points which are not important enough to form a class. However, the main weakness of DBSCAN is its computational complexity which is $\mathcal{O}(n^2)$, where n is the number of data points.

Many works have proposed approaches to scale-up the DBSCAN algorithm by parallelizing its execution [9, 15–17, 25, 26, 28, 31, 32]. These approaches propose to partition the dataset, to cluster the entities within the partition and to merge the clusters produced by each partition to provide the final result. They are parallelized and implemented using a big data technology; they achieve a fast clustering process. However, these approaches are executed in one batch and are not incremental. Thus, using these algorithms on a dataset which evolves over time would require to repeat their execution on the global dataset after each insertion.

Some approaches have proposed an incremental version of DBSCAN. In [10], the neighborhood of an inserted or deleted entity is computed and some rules are proposed in order to update the corresponding clusters. However, this approach processes one entity at a time. In addition, updating the clusters after the insertion of an entity requires its comparison with the entire dataset, which is a costly operation. [27] proposes to enhance the previous approach by limiting the search space to partitions during the neighbors computation, rather than the whole dataset. The dataset is split into partitions based on partition centers, and a new entity is assigned to the partition with the closest center. The neighborhood of the new entity is computed within this partition only. However, defining a center in an RDF dataset is not straightforward. In addition, partitioning data based on centers does not ensure that the result is the same as the one of the DBSCAN algorithm, which could decrease the quality of the clustering. RT-DBSCAN [13] proposes to define the $((minPts\text{-}1) \times \epsilon)$-neighborhood of the new inserted entity and to perform the clustering in this region using DBSCAN. It parallelizes the execution of the approach by dividing the dataset into cells where the incremental algorithm is executed in parallel, then the clusters produced for each cell are merged to build the final clustering result. This algorithm is implemented using Spark streaming. However, this approach is designed for data represented in a 2D space and is not suitable for RDF data.

7   Conclusion

In this work, we have addressed the problem of incrementally extracting a schema from large RDF datasets that evolve over time.

We have proposed a novel incremental density-based clustering algorithm that scales up to very large RDF datasets. It builds the clusters which group similar entities by updating the existing clusters according to the neighborhood of inserted or deleted entities, and ensures that the resulting set of clusters is the same as the one generated using DBSCAN on the global dataset. The clusters produced by our approach represent the classes of the schema that capture the structure of the entities contained in an RDF dataset.

The present work describes an incremental schema discovery approach that ensures updating the schema after the insertion or deletion of entities to keep it coherent with the data. Beside insertion or deletion, RDF data sources may also evolve by updating the properties describing an existing entity. In such case, all the cases identified for both the insertion and the deletion of entities may occur, including the creation or deletion of a cluster, the merging or split of existing cluster, extending or reducing of a cluster. Indeed, as the similarity between entities is based on their shared properties, updating the properties describing an existing entity changes the similarity values and as a consequence, changes its neighborhood in two possible ways. First, new entities could be added to the neighborhood of the updated entity which could lead to the creation of a new cluster, the extension of an existing cluster or the merging of clusters. This case is the same as schema evolution after entity insertion described in Sect. 3. Second, entities could be deleted from the neighborhood of the updated entity which could lead to the deletion, the reduction or the split of an existing cluster. This case is the same as schema evolution after the deletion of existing entities described in Sect. 4.

Our proposal has been implemented using Spark, a big data technology for distributed large-scale data processing which has enabled the clustering of large RDF datasets. The experiments have shown that incrementally extracting a schema from an RDF dataset using our approach outperforms the existing scalable schema discovery approach using DBSCAN when applied on the global dataset, with both synthetic and real data.

In our future works, we will explore the possible ways of enriching the set of classes provided by our approach, by generating the semantic links between these classes as well as providing some semantic annotations. Besides, as some schema-related declarations could be available in the dataset, another possible way of improving our approach is to extend our algorithms in order to exploit partially available schema-related declarations to guide the discovery process, which could improve significantly the quality of the resulting schema.

References

1. Alcalde, C., Burusco, A.: Study of the relevance of objects and attributes of L-fuzzy contexts using overlap indexes. In: Medina, J., et al. (eds.) IPMU 2018. CCIS, vol. 853, pp. 537–548. Springer, Cham (2018). https://doi.org/10.1007/978-3-319-91473-2_46
2. Auer, S., Bizer, C., Kobilarov, G., Lehmann, J., Cyganiak, R., Ives, Z.: DBpedia: a nucleus for a web of open data. In: Aberer, K., et al. (eds.) ASWC/ISWC -2007. LNCS, vol. 4825, pp. 722–735. Springer, Heidelberg (2007). https://doi.org/10.1007/978-3-540-76298-0_52
3. Baazizi, M.A., Lahmar, H.B., Colazzo, D., Ghelli, G., Sartiani, C.: Schema inference for massive JSON datasets. In: Proceeding of the 20th International Conference on Extending Database Technology (EDBT), pp. 222–233 (2017)
4. Baazizi, M.A., Lahmar, H.B., Colazzo, D., Ghelli, G., Sartiani, C.: Parametric schema inference for massive JSON datasets. VLDB J. **28**, 497–521 (2019)
5. Bouhamoum, R., Kedad, Z., Lopes, S.: Scalable schema discovery for RDF data. Trans. Large Scale Data Knowl. Centered Syst. **46**, 91–120 (2020). https://doi.org/10.1007/978-3-662-62386-2_4
6. Bouhamoum, R., Kedad, Z., Lopes, S.: Incremental schema discovery at scale for RDF data. In: Verborgh, R., et al. (eds.) ESWC 2021. LNCS, vol. 12731, pp. 195–211. Springer, Cham (2021). https://doi.org/10.1007/978-3-030-77385-4_12
7. Bouhamoum, R., Kellou-Menouer, K.K., Lopes, S., Kedad, Z.: Scaling up schema discovery approaches. In: Proceeding of the 34th International Conference on Data Engineering Workshops (ICDEW), pp. 84–89. IEEE (2018)
8. Christodoulou, K., Paton, N.W., Fernandes, A.A.A.: Structure inference for linked data sources using clustering. Trans. Large Scale Data Knowl. Centered Syst. **19**, 1–25 (2015). https://doi.org/10.1007/978-3-662-46562-2_1
9. Cordova, I., Moh, T.: DBSCAN on resilient distributed datasets. In: 2015 International Conference on High Performance Computing & Simulation, HPCS 2015, Amsterdam, Netherlands, 20–24 July 2015, pp. 531–540. IEEE (2015). https://doi.org/10.1109/HPCSim.2015.7237086
10. Ester, M., Kriegel, H., Sander, J., Wimmer, M., Xu, X.: Incremental clustering for mining in a data warehousing environment. In: Gupta, A., Shmueli, O., Widom, J. (eds.) VLDB 1998, Proceedings of 24rd International Conference on Very Large Data Bases, 24–27 August 1998, New York City, New York, USA, pp. 323–333. Morgan Kaufmann (1998). http://www.vldb.org/conf/1998/p323.pdf
11. Ester, M., Kriegel, H.P., Sander, J., Xu, X.: A density-based algorithm for discovering clusters in large spatial databases with noise. In: Proceeding of the Second International Conference on Knowledge Discovery and Data Mining (KDD), pp. 226–231. AAAI Press (1996)
12. The Apache Software Foundation: Apache Hadoop (2018). https://hadoop.apache.org/. Accessed 20 Oct 2018
13. Gong, Y., Sinnott, R.O., Rimba, P.: RT-DBSCAN: real-time parallel clustering of spatio-temporal data using spark-streaming. In: Shi, Y., et al. (eds.) ICCS 2018. LNCS, vol. 10860, pp. 524–539. Springer, Cham (2018). https://doi.org/10.1007/978-3-319-93698-7_40
14. Gragera Aguaza, A., Suppakitpaisarn, V.: Relaxed triangle inequality ratio of the Sørensen-Dice and Tversky indexes. Theor. Comput. Sci. **718**, 37–45 (2017)

15. Han, D., Agrawal, A., Liao, W., Choudhary, A.N.: A novel scalable DBSCAN algorithm with spark. In: 2016 IEEE International Parallel and Distributed Processing Symposium Workshops, IPDPS Workshops 2016, Chicago, IL, USA, 23–27 May 2016, pp. 1393–1402. IEEE Computer Society (2016). https://doi.org/10.1109/IPDPSW.2016.57

16. He, Y., Tan, H., Luo, W., Feng, S., Fan, J.: MR-DBSCAN: a scalable MapReduce-based DBSCAN algorithm for heavily skewed data. Front. Comp. Sci. **8**(1), 83–99 (2014). https://doi.org/10.1007/s11704-013-3158-3

17. He, Y., et al.: MR-DBSCAN: an efficient parallel density-based clustering algorithm using mapreduce. In: 17th IEEE International Conference on Parallel and Distributed Systems, ICPADS 2011, Tainan, Taiwan, 7–9 December 2011, pp. 473–480. IEEE Computer Society (2011). https://doi.org/10.1109/ICPADS.2011.83

18. IBM: IBM quest synthetic data generator (2015). https://sourceforge.net/projects/ibmquestdatagen/. Accessed 01 Oct 2018

19. Jaccard, P.: The distribution of flora in the alpine zone. New Phytol. **11**(2), 37–50 (1912)

20. Jafari, O., Maurya, P., Nagarkar, P., Islam, K.M., Crushev, C.: A survey on locality sensitive hashing algorithms and their applications. CoRR abs/2102.08942 (2021). https://arxiv.org/abs/2102.08942

21. Kardoulakis, N., Kellou-Menouer, K., Troullinou, G., Kedad, Z., Plexousakis, D., Kondylakis, H.: Hint: hybrid and incremental type discovery for large RDF data sources. In: Zhu, Q., Zhu, X., Tu, Y., Xu, Z., Kumar, A. (eds.) SSDBM 2021: 33rd International Conference on Scientific and Statistical Database Management, Tampa, FL, USA, 6–7 July 2021, pp. 97–108. ACM (2021). https://doi.org/10.1145/3468791.3468808

22. Kellou-Menouer, K., Kardoulakis, N., Troullinou, G., Kedad, Z., Plexousakis, D., Kondylakis, H.: A survey on semantic schema discovery. VLDB J. (2021). https://doi.org/10.1145/3468791.3468808

23. Kellou-Menouer, K., Kedad, Z.: Schema discovery in RDF data sources. In: Johannesson, P., Lee, M.L., Liddle, S.W., Opdahl, A.L., López, Ó.P. (eds.) ER 2015. LNCS, vol. 9381, pp. 481–495. Springer, Cham (2015). https://doi.org/10.1007/978-3-319-25264-3_36

24. Kellou-Menouer, K., Kedad, Z.: A self-adaptive and incremental approach for data profiling in the semantic web. Trans. Large Scale Data Knowl. Centered Syst. **29**, 108–133 (2016). https://doi.org/10.1007/978-3-662-54037-4_4

25. Lulli, A., Dell'Amico, M., Michiardi, P., Ricci, L.: NG-DBSCAN: scalable density-based clustering for arbitrary data. Proc. VLDB Endow. **10**(3), 157–168 (2016). https://doi.org/10.14778/3021924.3021932

26. Luo, G., Luo, X., Gooch, T.F., Tian, L., Qin, K.: A parallel DBSCAN algorithm based on spark. In: Cai, Z., et al. (eds.) 2016 IEEE International Conferences on Big Data and Cloud Computing (BDCloud), Social Computing and Networking (SocialCom), Sustainable Computing and Communications (SustainCom), BDCloud-SocialCom-SustainCom 2016, Atlanta, GA, USA, 8–10 October 2016, pp. 548–553. IEEE Computer Society (2016). https://doi.org/10.1109/BDCloud-SocialCom-SustainCom.2016.85

27. Bakr, A.M., Ghanem, N.M., Ismail, M.A.: Efficient incremental density-based algorithm for clustering large datasets. Alex. Eng. J. **54**, 1147–1154 (2015)

28. Patwary, M.M.A., Palsetia, D., Agrawal, A., Liao, W., Manne, F., Choudhary, A.N.: A new scalable parallel DBSCAN algorithm using the disjoint-set data structure. In: Hollingsworth, J.K. (ed.) SC Conference on High Performance Computing Networking, Storage and Analysis, SC 2012, Salt Lake City, UT, USA, 11–15 November 2012, p. 62. IEEE/ACM (2012). https://doi.org/10.1109/SC.2012.9
29. Pernelle, N., Saïs, F., Mercier, D., Thuraisamy, S.: RDF data evolution: efficient detection and semantic representation of changes. In: Proceedings of the Posters and Demos Track of the International Conference on Semantic Systems - SEMANTICS, vol. 12 (2016)
30. Sevilla Ruiz, D., Morales, S.F., García Molina, J.: Inferring versioned schemas from NoSQL databases and its applications. In: Johannesson, P., Lee, M.L., Liddle, S.W., Opdahl, A.L., López, Ó.P. (eds.) ER 2015. LNCS, vol. 9381, pp. 467–480. Springer, Cham (2015). https://doi.org/10.1007/978-3-319-25264-3_35
31. Savvas, I.K., Tselios, D.C.: Parallelizing DBSCAN algorithm using MPI. In: Reddy, S., Gaaloul, W. (eds.) 25th IEEE International Conference on Enabling Technologies: Infrastructure for Collaborative Enterprises, WETICE 2016, Paris, France, 13–15 June 2016, pp. 77–82. IEEE Computer Society (2016). https://doi.org/10.1109/WETICE.2016.26
32. Song, H., Lee, J.: RP-DBSCAN: a superfast parallel DBSCAN algorithm based on random partitioning. In: Das, G., Jermaine, C.M., Bernstein, P.A. (eds.) Proceedings of the 2018 International Conference on Management of Data, SIGMOD Conference 2018, Houston, TX, USA, 10–15 June 2018, pp. 1173–1187. ACM (2018). https://doi.org/10.1145/3183713.3196887
33. Issa, S., Paris, P.-H., Hamdi, F., Si-Said Cherfi, S.: Revealing the conceptual schemas of RDF datasets. In: Giorgini, P., Weber, B. (eds.) CAiSE 2019. LNCS, vol. 11483, pp. 312–327. Springer, Cham (2019). https://doi.org/10.1007/978-3-030-21290-2_20
34. The Apache Software Foundation: Apache Spark (2018). https://spark.apache.org. Accessed 20 Oct 2018

Optimizing Data Coverage
and Significance in Multiple Hypothesis
Testing on User Groups

Nassim Bouarour$^{(\boxtimes)}$, Idir Benouaret, and Sihem Amer-Yahia

CNRS, Univ. Grenoble Alpes, Grenoble, France
{nassim.bouarour,idir.benouaret,sihem.amer-yahia}@univ-grenoble-alpes.fr

Abstract. We tackle the question of checking hypotheses on user data. In particular, we address the challenges that arise in the context of testing an input hypothesis on many data samples, in our case, user groups. This is referred to as Multiple Hypothesis Testing, a method of choice for data-driven discoveries. Ensuring sound discoveries in large datasets poses two challenges: the likelihood of accepting a hypothesis by chance, i.e., returning false discoveries, and the pitfall of not being representative of the input (data coverage). We develop GROUPTEST, a framework for group testing that addresses both challenges. We formulate VALMIN and COVMAX, two generic top-n problems that seek n user groups satisfying one-sample, two-sample, or multiple-sample tests. VALMIN optimizes significance while setting a constraint on data coverage and COVMAX aims to maximize data coverage while controlling significance. We show the hardness of VALMIN and COVMAX. We develop a greedy algorithm to solve the former problem and two algorithms to solve the latter where the first one is a greedy algorithm with a provable approximation guarantee and the second one is a heuristic-based algorithm based on α-investing. Our extensive experiments on real-world datasets demonstrate the necessity to optimize coverage for sound discoveries on large datasets, and the efficiency of our algorithms.

Keywords: Hypothesis testing · Data coverage · Exploratory data analysis

1 Introduction

Understanding people and their preferences in the Internet of Behaviors[1] requires expressive and sound methods for data-driven discoveries. Making sound discoveries in large datasets poses two challenges: the likelihood of accepting a hypothesis by chance, i.e., returning false discoveries, and the pitfall of not being representative of the input, i.e., data coverage. In this paper, we develop GROUPTEST,

---

[1] https://www.gartner.com/smarterwithgartner/gartner-top-strategic-technology-trends-for-2021/.

© Springer-Verlag GmbH Germany, part of Springer Nature 2022
A. Hameurlain et al. (Eds.): *Transactions on Large-Scale Data- and Knowledge-Centered Systems LI*, LNCS 13410, pp. 64–96, 2022.
https://doi.org/10.1007/978-3-662-66111-6_3

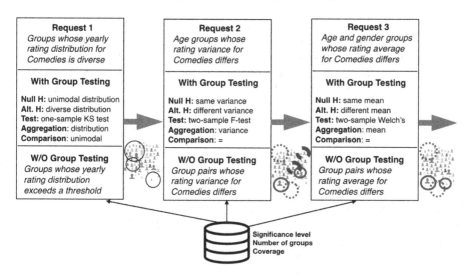

Fig. 1. A multi-step group testing.

a framework for group testing that seeks to find representative and statistically significant user groups. GROUPTEST has several applications ranging from verifying hypotheses on patient data to the social Web.

A statistical hypothesis test compares two models (the null and the alternative hypotheses) and deems the comparison statistically significant if, according to a significance threshold α, the data is very unlikely to have occurred under the null hypothesis, i.e., the null hypothesis is rejected, and the alternative hypothesis is satisfied. To leverage hypothesis testing in modern applications, existing work proposed to combine multiple hypothesis testing with data mining [14,25,28], data exploration [32] or machine learning [29]. The main focus of those works is to control the risk of false discoveries when a large number of hypotheses have to be tested. The higher the number of hypotheses that are tested, the higher the probability of making a false discovery [16]. This requires to use p-value correction methods [13,20] such as the Bonferroni Family-Wise Error Rate (FWER) or the Benjamini-Yekutieli False Discovery Rate (FDR) [5]. Additionally, in a context where an analyst needs to take actionable insights, a limited number of results that are both valid with respect to a statistical test and representative of the data, is desirable but the sole focus on hypothesis significance may lead to exploring only a small portion of the data of interest. To remedy that, we propose to combine data coverage with significance correction methods.

Motivating Example 1. Consider an analyst who seeks to examine movie ratings with the goal of forming an international panel of diverse experts to judge Comedies. In this case, data samples refer to reviewer groups. Figure 1 illustrates the example in 3 steps. **Request 1** looks for reviewer groups who provided diverse ratings for Comedies. A one-sample Kolmogorov-Smirnov test identifies three groups that reject the null hypothesis "Rating distribution is unimodal" (and satisfy the alternative hypothesis stated in the request). An additional constraint is needed to ensure that returned groups are representative of the input, i.e. cover most reviewers of Comedies. **Request 2** further refines returned groups by exploring their age subgroups. It applies a two-sample F-test that seeks to reject the null hypothesis "Age groups whose variance for Comedies is the same". Here again, traditional hypothesis testing will only check rating variances. An additional constraint is needed to ensure that the input age groups are covered by the resulting groups. The output groups are fed to a two-sample Welsh test to compare their average ratings in **Request 3**, and return those that significantly differ. Members of those groups can be used by the analyst to form the desired panel.

Motivating Example 2. Consider an analyst who wants to search pairs of groups satisfying the following request: "Female groups whose rating mean is lower than Male groups". The analyst limits the number of returned results to $n = 3$. The analyst chooses the most significant pairs, represented in the Table 1. The two first female and male groups are overlapping. Indeed, most under 18 yo female users are High school students while on the other hand, practically the same group of males is returned ([50–55] aged users). The returned pairs of groups do not optimize coverage as they cover no more than 22% of data.

Table 1. Top-3 most significant results that satisfy the request "Female groups whose rating mean is lower than Male groups".

Female groups	Male groups
Under 18 aged users who rated 90's movies	[50–55] aged users who rated 90's movies
High school students	[50–55] aged users who rated long movies
[25–34] aged users who rated Drama movies	[35–44] aged technical engineers

After that, the analyst wants to search pairs satisfying the same request, but this time, chooses to seek for pairs that maximize the coverage of data. The returned pairs are presented in the Table 2. The female and male groups are larger and compact (with less descriptive attributes) than the previous ones. Which results to pairs that maximize coverage and with a minimum overlap. The analyst found pairs that have a slightly low significance score than the previous ones but cover more than 91% of the data.

Table 2. Top-3 of results that cover the most the data and satisfy the request "Female groups whose rating mean is lower than Male groups".

Female groups	Male groups
[25–34] aged users	Users who rated Comedy movies
Users who rated Drama movies	Users who rated 90's movies
Users who rated Romance movies	Users who rated 2000's movies

Challenges. Realizing our examples requires to address two challenges: (i) the likelihood of rejecting a null hypothesis and returning groups by chance, when the number of groups grows, and (ii) the pitfall of returning groups that are not representative of the input data of interest. Indeed, when multiple data samples are tested against a hypothesis, the chance of observing a rare event increases, and hence, the likelihood of incorrectly rejecting a null hypothesis (i.e., making a Type I error [26]) increases. There are precise criteria for excluding or not a null hypothesis at a certain significance level [16,20]. Those criteria depend on the type of test (i.e., one-sample, two-sample, multiple-sample), the aggregation function (e.g., mean, variance), the sample sizes, whether the samples are paired (same subjects), and the comparison operators (e.g., equal, greater). The second challenge is particularly important for large datasets where the number of groups that pass the test increases and the choice of which ones to return may affect how representative they are, i.e. how much they cover the input. This requires to revisit the conditions under which a null hypothesis is rejected to additionally account for data coverage when selecting groups to return.

Contributions. We develop GROUPTEST, a unified framework that supports a variety of statistical tests to verify if user behavior supports the null or the alternative hypotheses, and return qualifying groups. We, first, formulate a generic problem VALMIN as an extension of the traditional multiple hypothesis testing problem. VALMIN is a top-n problem that seeks n user groups accommodating different hypothesis tests (one-sample, two-sample, or multiple-sample tests) and additionally satisfying a constraint on data coverage. This problem enforces coverage but does not maximize it. We hence propose to generalize it by formulating CoVMAX which is also a generic top-n problem that maximizes data coverage while setting a constraint on hypothesis significance. VALMIN and CoVMAX leverage different hypothesis correction methods, FWER and FDR.

We show that both problems are NP-hard with a reduction to the Partial Weighted Set Cover Problem and the Maximum Coverage Problem respectively. We develop VAL_C, a greedy algorithm that solves VALMIN, and COVER_G, a greedy algorithm that solves CoVMAX. To address scalability, we also develop COVER_α, a more efficient heuristic algorithm to solve CoVMAX.

VAL_C iterates over the set of all candidate groups and chooses the ones that maximize significance (smallest p-values). Similarly to traditional procedures, VAL_C controls the multiple testing error at a given significance level α (usually set to 0.05). To do that, it calculates and ranks all candidates on decreasing p-value. VAL_C iterates over the candidates while checking the significance of each new candidate and the satisfaction of the coverage constraint. To achieve that and return n results, VAL_C greedily replaces the previously selected groups with new candidates until the coverage constraint is satisfied.

COVER_G is a greedy algorithm, with a provable approximation guarantee. As VAL_C, it iterates over the set of candidate groups and instead of choosing the candidates that minimize significance, it chooses the next candidate that maximizes coverage. It also controls significance by calculating and ranking the p-values of all candidates. This hinders its scalability when the number of groups increases. To address that, we propose COVER_α, a heuristic algorithm that builds on α-investing [11], an adaptive sequential method that controls mFDR, the ratio of the expected number of false rejections to the expected number of rejections. Different investing policies of the α-wealth have been proposed previously albeit without considering data coverage [32]. The key idea of COVER_α is to invest more significance, referred to as α-wealth, on candidates with the highest coverage. This decision relies on tuning a hyper parameter λ whose value determines the speed at which the α-wealth is consumed, and needs to be explored empirically.

Experiments. We run extensive experiments on multiple datasets to explore the interplay between hypothesis significance and data coverage. We report results on **MovieLens'1M** and refer to our Github repository[2] for other datasets.

For the VALMIN problem, we compare VAL_C against an adaptation of a Subfamily wise procedure (SMT) [29]. We compare power (#true positives/#results in ground-truth) and FDR (#false positives/#results) and find that satisfying coverage comes at a negligible significance cost. Indeed, for small n ($n \leq 50$), SMT slightly outperforms VAL_C in terms of significance without guaranteeing on coverage satisfaction. For higher n values, VAL_C achieves the same significance as SMT while also satisfying coverage. For the COVMAX problem, we compare COVER_G and COVER_α against traditional correction methods as well as different α-investing policies [32]. We find that COVER_α performs better than COVER_G and other α-investing policies in terms of significance. We also examine how varying the sample size and the number of results affects coverage. We find that COVER_α yields very good results as it performs closely to COVER_G in terms of coverage while having a much smaller response time.

This demonstrates that our solutions attain high coverage values while ensuring sound group testing in reasonable times.

Organization. Section 2 reviews the related work. Section 3 introduces the GROUPTEST framework and presents our problems. Section 4 contains our algorithms. Experiments are described in Sect. 5. We conclude the paper in Sect. 6.

2   Related Work

Similarly to existing work on combining data mining, data exploration, visualization or machine learning with hypothesis testing, we leverage hypothesis test correction methods to control the risk of false discoveries on user groups.

Existing work on customer segment discovery [14,25] combines the computational power of pattern mining with multiple hypothesis testing to find meaningful patterns in the data. The first step is to find genuine patterns that are likely to reflect properties of the underlying population and hold also in the data samples. A variety of statistical tests have been used to filter out patterns that are unlikely to be useful, removing uninformative variations of key patterns in the data [28]. Existing FWER and FDR methods are used to control the risk of false discoveries, i.e., of finding patterns that do not hold in the entire population. Zgraggen

---

[2] https://github.com/statistical-group-testing/statistically-soundgrouping.

et al. [31] conducted a user study about visual inference and randomness in multiple testing on visualization systems using synthetic datasets. They concluded that without statistical verification, more than 60% of the insights reported by users were false. They also showed that multiple testing significantly decreases FDR and compared it to different hypotheses confirmation techniques.

The idea of correcting for a cascade of hypotheses, i.e., verifying hypotheses on multiple data samples, was considered in [29]. The authors introduce Subfamily-wise Multiple Testing, a multiple-testing correction that can be used when there are repeated pools of hypotheses from each of which a single null hypothesis is to be rejected and neither the specific hypotheses nor their number are known until the final rejection decision. A Monte Carlo algorithm is used to apply multiple testing correction to a cascade of statistical hypothesis tests subject to a risk budget. The authors show how this can be applied to select graphical models in machine learning. A similar idea is considered in the context of interactive data exploration where a series of hypotheses are verified [32], where each hypothesis corresponds to an exploration action. The authors develop a set of control procedures that are tailored to interactive data exploration. In particular, they leverage the α-investing technique proposed in [11] in their heuristics and show that they can better control errors and increase the power of hypothesis testing in interactive data exploration.

Existing work has also combined machine learning and statistical testing to verify associations in genome data (e.g. [21]). The proposed algorithm first trains a support vector machine to determine a subset of candidates and then performs hypothesis tests using a correction threshold. The authors show that by taking into account correlation structures, their method outperforms ordinary raw p-value thresholding. Their solution is shown to yield fewer false (i.e. non-replicated) and more true (i.e. replicated) discoveries.

The difference between previous works and ours is that none of them addresses data coverage explicitly. To the best of our knowledge, we are the first to adopt a data-centric approach to multi-hypothesis testing. Our formalization seeks to return groups that maximize coverage, i.e., that are representative of the input data. Our GROUPTEST framework accommodates various correction methods and examines the interplay between hypothesis testing and data coverage. Its appeal also lies in its generic nature as it handles different test types and aggregations (mean, etc.), whether the samples are paired or not, and different comparison operators. Our work is also different as we are the first to perform scalability tests in multi-hypothesis testing.

3   The GROUPTEST Framework

We first present motivating examples. Our framework is applicable to data modeled as a bipartite graph formed by users and items with their respective attributes (See Sect. 3.2). This model is generic enough to capture many datasets on the social Web. We end this section by formulating our problems.

3.1 Motivating Examples

Our purpose is to develop a powerful framework for group testing. A one-sample, two-sample or multiple-sample hypothesis is verified when *groups* identified by some *filters*, have statistically *similar, higher, lower aggregates* with respect to some aggregate (mean, variance, rating distribution on group members). This is referred to as the alternative hypothesis that states the desired test and complements the null hypothesis. Therefore, the null hypothesis is said to be rejected by desired groups.

Table 3 illustrates the variety of requests we handle in GROUPTEST with examples on movie ratings along with the type of test that is relevant for each request (third column). The first four types of requests use mean to aggregate group ratings and require different types of tests. R_1 shows the case of a one-sample t-test. Input data is all movie ratings by students. Subgroups such as "Students in California" or "Young students" are generated and their average rating is compared to a reference value (here, 3.5) with a one-sample test. Groups that reject the null hypothesis "Students whose rating average is equal to 3.5" and satisfy the alternative hypothesis "Students whose rating average is greater than 3.5" are returned. The case of a two-sample test is shown in R_2 and R_3. In R_4, we compare groups across 3 weeks and return tuples of groups whose rating mean for 80's movies differs. We use variance as an aggregation in R_6 and rely on a two-sample F-test. It starts with all rating records for movies in the 70's and returns pairs of groups whose rating variance for those movies differs in the Spring. The last two types, R_7 and R_8, compare rating distributions using one-sample and two-sample Kolmogorov-Smirnov tests.

3.2 Groups

Given a set of users $\mathcal{U}$ and a set of items $\mathcal{I}$, we define *user data* as a database $\mathcal{D}$ of tuples $\langle u, i, a \rangle$ where $a \in \mathbb{R}$ is a value induced by an action such as browsing, tagging, or rating, of user $u \in \mathcal{U}$, on item $i \in \mathcal{I}$. Users have attributes drawn from a set $A_{\mathcal{U}}$ and items have attributes drawn from a set $A_{\mathcal{I}}$. For example, users can be represented with $A_{\mathcal{U}} = \langle$ uid, age, gender, occupation, location $\rangle$, a user instance may be $\langle 568, 18$–$24,$ female, student, NY $\rangle$. Similarly, items on Movielens can be represented with $A_{\mathcal{I}} = \langle$ title, genre, year, run time $\rangle$ and the movie *Titanic* as $\langle$ Titanic, Romance & Drama, 1997, 195 $\rangle$.

Definition 1 (Group). *A group g is a set of records where users have at least one common attribute value (e.g., same gender) and may have some common actions (e.g., rated the same movies).*

For instance, $g = [\langle$ gender, female $\rangle, \langle$ age, 25–34 $\rangle, \langle$ item title, Titanic $\rangle]$ contains 25–34 aged females who rated the movie Titanic.

We use $\mathcal{G}$ to denote the set of all groups. Hence, $|\mathcal{G}|$ is the powerset of user and item attribute values. For instance, with $|A_{\mathcal{U}}| = 4$ and 3 values per attribute, and $|A_{\mathcal{I}}| = 3$ with 5 values, $|\mathcal{G}| = 2^{(4 \times 3 + 3 \times 5)}$.

Table 3. Examples of group testing requests in GROUPTEST with "groups", **aggregate**, dimension, and *operator* with the corresponding statistical test.

R₁	"Student groups" whose <u>rating</u> **mean** *is greater than 3.5*	One-sample t-test
R₂	"Female groups" whose <u>rating</u> **mean** *is lower than* "Male groups" within the same period	Two-sample Welch's test
R₃	"Male Groups" whose <u>rating</u> **mean** *changes* between 2 Seasons	Two-sample paired t-test
R₄	"Groups" whose <u>rating</u> **mean** for "80's movies" *differs* in a 3-week period	Multiple mean F-test: ANOVA
R₅	"Groups" whose <u>rating</u> **variance** for "Comedy movies" *is greater than 1*	One-sample variance Chi-square test
R₆	"Group pairs" whose <u>rating</u> **variance** for "70's movies" *differs* in the Spring	Two-sample variance F-test
R₇	"Groups" whose yearly <u>rating</u> **distribution** *does not follow a Gaussian distribution*	One-sample Kolmogorov-Smirnov test
R₈	"Group pairs" whose <u>rating</u> **distribution** for "Drama movies" *differs* in the same season	Two-sample Kolmogorov-Smirnov test

In the literature, groups have been referred to with different terms, such as *communities* [22], *tribes* [12], *cliques* [7], *cohorts* [17], *teams* [23], *segments* [3], *patterns* [6,30], *cubes* [18], *clusters* [2,27] and *partitions* [24]. Our model is designed to be agnostic about the approach used to compute groups.

3.3   Group Testing

A hypothesis test considers two hypotheses that contain opposing viewpoints. The null hypothesis H_0 usually states that group aggregates are the same. The alternative hypothesis H_a states a claim that contradicts H_0 and corresponds to desired samples (in our case, user groups). The decision can either be "reject H_0" if the sample favors the alternative hypothesis or "do not reject H_0" if the sample is insufficient to reject the null hypothesis.

The GROUPTEST framework is aimed to be generic and accommodates various types of tests. Different statistical tests qualify depending on group size, group members (paired or unpaired), and the aggregation function AGG. Figure 2 summarizes the aggregation functions and statistical tests considered in our work. The last column refers to the requests shown in Table 3.

AGGREGATE	INFERENCE and TEST	DEFINITION and Example	
Mean	Test about a mean: One-sample t-test	$t = \dfrac{\bar{x} - \mu_0}{s/\sqrt{n}}$ with $\bar{x}$ and s, the sample mean and standard deviation, and μ_0 the reference mean, and n the sample size	R1
	Test to compare two means: Two-sample Welch's t-test	$t = \dfrac{(\bar{x}_1 - \bar{x}_2)}{\eta}$ with $\bar{x}_1 - \bar{x}_2$ the difference between 2 sample means, s the pooled standard deviation, and $\eta = \begin{cases} s\sqrt{2/n}, & \text{for } n = n_1 = n_2 \\ s\sqrt{1/n_1 + 1/n_2}, & \text{for } n_1 \neq n_2 \text{ and } s = s_1 = s_2 \\ \sqrt{s_1/n_1 + s_2/n_2}, & \text{for } n_1 \neq n_2 \text{ and } s_1 \neq s_2 \end{cases}$	R2
	Test about a mean with paired data: Paired difference t-test	$t = \dfrac{\bar{d} - d_0}{s_d/\sqrt{n}}$ with $\bar{d}$ and s_d the average and standard deviation of the differences all pairs and the reference difference d_0	R3
	Test to compare K multiple means: F-test for one way ANOVA	$F = \dfrac{MST}{MSE}$ with $MST = \dfrac{\sum_{i=1}^{K} n_i(\bar{x}_i - \bar{x})^2}{K-1}$, $MSE = \dfrac{\sum_{i=1}^{K}(n_i - 1)s_i^2}{n - K}$ and $n = n_1 + \cdots + n_K$, $\bar{x} = \dfrac{\sum_{i=1}^{K} x_i}{n}$	R4
Variance	Test about a population variance: Chi-squared test	$T = (n-1)\, s^2/\sigma_0^2$ with s^2 the sample variance, n the sample size and σ^2 the reference variance	R5
	Test to compare two population variances: F-test	$F = s_1^2/s_2^2$ with s_1^2 and s_2^2 the sample variances of the 2 populations	R6
Distribution	Test about a distribution: One-sample Kolmogorov–Smirnov test	$D_n = sup_x |F_n(x) - F_0(x)|$ with F_n the empirical distribution function, F the reference distribution and sup the supremum function	R7
	Test to compare two distributions: Two-sample Kolmogorov-Smirnov test	$D_{n,m} = sup_x |F_{1,n}(x) - F_{2,m}(x)|$ with $F_{1,n}, F_{2,m}$ the empirical distribution functions of the two samples and sup is the supremum function	R8

Fig. 2. Summary of statistical tests considered in GROUPTEST.

Definition 2 (Group testing request). *A group testing request R is a tuple $\langle H_0, H_a, \mathrm{MSR}, \mathrm{AGG}, \mathrm{OP} \rangle$ where H_0 is a null hypothesis, H_a an alternative hypothesis, MSR is a user behavior dimension (e.g., rating, purchase), AGG is an aggregation function applied to a behavior dimension (average, variance, distribution), and OP the operator used to compare aggregates $(=, <>, >, \text{ and } <)$.*

To simplify our notation, we omit the condition on user and item attributes in R and assume that a request R is evaluated on $D \subseteq \mathcal{D}$ where those conditions are satisfied. The subset D is used to create a set of groups $G \subseteq \mathcal{G}$. To evaluate a request R, we compute a set of *allCandidates* as follows: for one-sample tests, *allCandidates* $= \{\langle g \rangle\}$; for two-sample tests, *allCandidates* $= \{\langle g, g' \rangle\}$; for multiple-sample tests *allCandidates* $= \{\langle g, g', \ldots \rangle\}$ where $g \in G, g' \in G, g <> g'$. We now describe how to compute the significance of each sample in *allCandidates* before we formalize our top-n problems.

Computing p-Values. The common protocol to compute p-values of each sample in *allCandidates* must first verify normality and independence of each sample [9]. Without loss of generality, we describe that protocol *for comparing two means with a two-sample t-test*. A value of 0.05 for α is commonly adopted and indicates a 5% risk of concluding that a difference exists between the two means when there is no actual difference [10]:

P-value Computation Protocol:
1. **Normality check:** Given a candidate pair $(g, g') \in$ *allCandidates*, verify that the data distribution of each group g and g' is normal, normalize it otherwise;
2. **Independence filtering:** Verify that the distributions of g and g' are independent using χ^2 test; Keep independent pairs;
3. **P-value computation:** Compute the p-value *pval* of independent (g, g') pairs wrt a request R.

3.4   Our Problems

In the following problems, the set *Candidates* contains the tuples $(g, g', pval)$ in D that passed p-value computation protocol wrt the hypothesis in R ($pval \leq \alpha$), *c.groups* denotes the group g, the pair (g, g') and the tuple $(g, g', ...)$ in c in the case of one-sample, two-sample and multiple-sample tests respectively. $cover(g, D)$ is defined as the intersection between users in group g and all users in dataset D. Formally, for a given group g, $cover(g, D) = g.users \cap D.users$. Moreover, *cover* is defined in the same way for all statistical tests and determined by the intersection between the union of all users in the groups *c.groups* and the users in the dataset D. For example, a coverage of a two-sample set of candidates $C = \{c_1, c_2\}$ is:

$$
\begin{aligned}
cover(\bigcup_{c.groups \in C}, D) &= cover(c_1.groups \cup c_2.groups, D) \\
&= cover((g_1 \cup g_1') \cup (g_2 \cup g_2'), D) \\
&= ((g_1.users \cup g_1'.users) \cup (g_2.users \cup g_2'.users)) \cap D.users
\end{aligned}
\tag{1}
$$

The definition of *Cover* for a larger set of candidates is straightforward

*Problem 1 (*VALMIN *Problem).* Given a request R, a dataset $D \subseteq \mathcal{D}$ that satisfies user and item conditions, a significance threshold θ on p-values, a minimum coverage value cov_{min}, a maximum number of desired results n, find a set C s.t.

$$C = \underset{C \subseteq Candidates}{\operatorname{argmin}} \sum_{c \in C} c.pval$$

subject to

$$|C| \leq n, \tag{2}$$
$$\forall c \in C, c.pval \leq \theta,$$
$$|cover(\bigcup_{c.groups \in C}, D)| \geq cov_{min}$$

The VALMIN Problem extends the multiple hypothesis testing by enforcing data representativity and having a constraint on coverage but it does not assure its maximization. We propose the CovMAX Problem to face this limitation and capture data coverage.

Problem 2 (CovMAX Problem). Given a request R, a dataset $D \subseteq \mathcal{D}$ that satisfies user and item conditions, a significance threshold θ on p-values, a maximum number of desired results n, find a set C s.t.

$$C = \underset{C \subseteq Candidates}{\operatorname{argmax}} |cover(\bigcup_{c.groups \in C}, D)|$$

subject to $\tag{3}$

$$|C| \leq n,$$
$$\forall c \in C \ c.pval \leq \theta$$

As the number of candidates increases, the likelihood that spurious hypotheses pass the test increases, causing Type I errors [26]. The significance level of p-values can be adjusted to control the expected proportion of incorrectly rejected null hypotheses. The simplest way to do so is to use the conservative Bonferroni correction [4], a Family-Wise Error Rate (FWER) control method. Bonferroni is preferred when false discoveries are not acceptable (in particular for critical decision-making, e.g., accepting a new medical treatment) or when it is expected that most null hypotheses would be true. A more powerful adjustment method is the Benjamini-Yekutieli False Discovery Rate (FDR) procedure [5] that allows to control the expected proportion of incorrectly rejected null hypotheses. FDR control is preferred in exploratory research, where the number of potential hypotheses is large and false discoveries are not so critical [14].

Our problem formulation is generic and aims to accommodate existing significance adjustment procedures by adapting the definition of the significance threshold θ as follows:

– For Bonferroni (BN): $\theta = \frac{\alpha}{m}$

– For Benjamini-Yekutieli (BY): $\theta = \frac{\alpha \times k}{m} \left(\sum_{i=1}^{m} 1/i \right)^{-1}$, where

$k = max \left\{ i : p_i \leq \frac{\alpha \times i}{m \times \sum_{i=1}^{m} 1/i} \right\}$, p_i the ith smallest p-value.

The value m is the number of groups in *Candidates* and α is the significance level usually set to 0.05 [10].

The drawback of these traditional adjustment methods is that to control FDR (or FWER) at a given level α, one has to previously calculate the p-values of all m candidates and then rank them to determine the correct threshold θ for rejecting the null hypothesis. This has two main limitations: (1) the calculation of p-values for all candidate groups is expensive and (2) there could be settings where we do not have a prior knowledge on the number of hypothesis m. To overcome this, we leverage the marginal FDR (*mFDR*) that was defined by Foster and Stine [11] that computes the ratio of the expected number of false rejections to the expected number of rejections as follows:

$$mFDR_\eta(j) = \frac{E[V(j)])}{E([R(j)]) + \eta} \tag{4}$$

where $V(j)$ designates the number of false discoveries (wrongly rejected null hypothesis) and $R(j)$ the total number of discoveries. The parameter η is used to weigh the impact of cases for which the number of discoveries is small, and η is usually set to 1 or $(1-\alpha)$ [32]. We can now reformulate the CovMax problem as follows:

Problem 3 (CovMax Problem reformulation). Given a request R, a dataset $D \subseteq \mathcal{D}$ that satisfies user and item conditions, a significance level α, a parameter η, a maximum number of desired results n, find a set C s.t.

$$C = \underset{C \subseteq Candidates}{\mathbf{argmax}} \; |cover(\bigcup_{c.groups \in C}, D)|$$

subject to
$$|C| \le n, \tag{5}$$
$$mFDR_\eta(j) \le \alpha$$

where j denotes the total number of hypothesis tests that have been performed.

Theorem 1. VALMIN *is NP-hard.*

Proof (sketch). Given $S = \{S_1, S_2, ..., S_m\}$ a collection of m sets where each set S_i is a subset of D which represents the set of all data elements. We assign to each set S_i a weight p_i. We want to identify the n sets that minimize the total weight and whose union covers a fraction β of D. This is known as the Partial Weighted Set Cover Problem [8] and is formulated as following:

$$C = \underset{i \subseteq \{1,..,m\}}{\mathbf{argmin}} \sum p_i$$

subject to
$$|C| \le n, \tag{6}$$
$$\frac{|\bigcup_{i \in C} S_i|}{|D|} \ge \beta$$

This problem is equivalent to the VALMIN when we add the significance constraint and we correspond each set S_i to the coverage of a candidate c_i from a set of m candidates, $S_i = cover(\cup_{c_i.groups}, D)$, each weight p_i to the p-value of the test of the candidate c_i, and the fraction β to the minimum coverage value cov_{min}.

The Partial Weighted Set Cover Problem represents a generalization of the Weighted Set Cover Problem [8] which is proved to be NP-hard [19]. This is sufficient to prove that our VALMIN problem is NP-hard.

Theorem 2. COVMAX *is NP-hard.*

Proof (sketch). Given $S = \{S_1, S_2, ..., S_m\}$ a collection of m sets where each set S_i is a subset of D which represents the set of all data elements. We want to identify the n sets that maximize the total coverage of D. This is known as the Maximum Coverage Problem and is formulated as following:

$$C = \operatorname*{argmax}_{i \subseteq \{1,..,m\}} \left| \bigcup_{i \in C} S_i \right|$$

subject to

$$|C| \leq n,$$

(7)

This problem is equivalent to the COVMAX when we add the significance constraint and we correspond each set S_i to the coverage of a candidate c_i from a set of m candidates, $S_i = cover(\cup_{c_i.groups}, D)$.

The Maximum Coverage Problem is proved to be NP-hard [15] which makes it sufficient to prove by reduction that our COVMAX problem is NP-hard.

4   Algorithms

Before we dive into our algorithms, we first describe how the set of candidate groups *Candidates* is generated. Without loss of generality, we illustrate our algorithms in the case of a request that requires two-sample tests. We define a sub-routine *GenerateCandidates* that takes a request R, a dataset D and generates the set of all groups *allCandidates*. For example, for R_2 in Table 3, it generates all groups that share the attribute value *female* and all groups that share the attribute value *male*, and creates *allCandidates* that contains pairs of groups (g, g') formed by a Cartesian product between the two sets. After that, another sub-routine *ComputePvalues* computes the p-value of each pair (g, g') in *allCandidates*, discards all pairs that have a p-value above the significance level α, and outputs a set *Candidates* of pairs along with their p-values.

4.1 Algorithm VAL_C

Algorithm 1: Minimum coverage algorithm (VAL_C) – illustrated with the Benjamini-Yekutieli correction

---

Input: a request R, a dataset D, a significance level α, a minimum coverage value cov_{min}, a number of results n

Output: C

1   $allCandidates \leftarrow GenerateCandidates(R, D)$
2   $Candidates \leftarrow ComputePvalues(allCandidates, \alpha)$
3   $C \leftarrow \emptyset$
4   $m \leftarrow |Candidates|$
5   $L = Sortbypval\ (Candidates)$
6   $k = \text{argmax}_{0 \leq j \leq m}\ P[j] \leq \frac{\alpha \times j}{m} \left(\sum\limits_{i=1}^{m} 1/i \right)^{-1}$
7   $C \leftarrow Top\text{-}n(L)$
8   $cov_C \leftarrow |cover(C.groups, D)|$
9   $L \leftarrow L \setminus C$
10   $i \leftarrow n + 1$
11   **while** $cov_C < cov_{min}$ *and* $i \leq k$ **do**
12     $c^* \leftarrow L[i]$
13     $i \leftarrow i + 1;\ L \leftarrow L \setminus \{c^*\}$
14     $C_{future} \leftarrow C$
15   **for** *each group* $\in C$ **do**
16       $C^* \leftarrow Swap(C, group, c^*)$
17   **if** $|cover(C^*.groups, D)| > cov_C$ **then**
18         $cov_C \leftarrow |cover(C^*.groups, D)|$
19         $C_{future} \leftarrow C^*$
20     $C \leftarrow C_{future}$
21   **return** C

To solve VALMIN, we propose a greedy algorithm VAL_C (Algorithm 1). It first generates *allCandidates* by calling the routine *GenerateCandidates()* (Line 1), computes their p-values (Line 2) and keeps the ones having significant p-values (*Candidates*). Then, it sorts the candidates by increasing order of their p-values into a list L (Line 5), before applying the significance adjustment procedure (Line 6). After that, it picks the set C of n candidates that have the smallest p-values and that are below the significance threshold $P[k]$ (Line 7), calculates their coverage (Line 8) and removes them from the set of candidates L (Line 9). If the selected groups C do not satisfy the minimum coverage constraint (cov_{min}) (Line 11), the algorithm will greedily scan the next candidates and at each step it will swap the pre-selected group in C that contributes the least to coverage with the considered candidate. Formally, VAL_C iterates through the remaining candidates (Line 12–13). The next candidate c^* is used to maximize the coverage in an iterative way (Line 15) where at each iteration a single pre-selected group from C is replaced by c^* generating by that a new set of groups C^* (Line 16). If the new set brings more coverage (Line 17), the procedure takes it as a potential future set of results (Line 19). The best set (the one that maximizes the most data coverage) is used to replace C (Line 20), otherwise C remains unchanged. The procedure stops either if the minimal coverage is reached or if all significant candidates were scanned.

The worst-case complexity of VAL_C is $O(m \cdot p + m \cdot log\ m + (k - n) \cdot n)$. The first term $m \cdot p$ represents the complexity of calculating the p-values of all m candidates by assuming that p is the worst-case complexity for computing a single p-value. The term $m \cdot log\ m$ is for sorting the candidates by their p-values and the term $(k-n) \cdot n$ is the complexity of scanning the remaining candidates and optimizing coverage. $k - n$ represents the number of the remaining candidates (k being the highest number of hypotheses that satisfy the constraint of multiple testing) and n represents the number of swaps performed for each candidate (Line 15–16 in Algorithm 1). We note that the number of candidates m is equal to the power set of attributes in the worst case.

4.2   Algorithm COVER_G

To solve CovMax, we propose COVER_G (Algorithm 2), a greedy algorithm. It first generates *allCandidates* by calling *GenerateCandidates()* (Line 1), computes their p-values and discards the non significant ones (Line 2), and then it sorts the candidates by increasing p-values into a list L (Line 3). The next step (Line 6) applies the significance adjustment procedure. We illustrate it here with Benjamini-Yekutieli. In this case, the algorithm calculates the greatest number k for which $L[j] \leq \frac{\alpha \times j}{m} \left(\sum_{i=1}^{m} 1/i \right)^{-1}$ is verified. After that, it greedily picks the next candidate that maximizes data coverage (Line 8) and removes it from the set of candidates (Line 9). It adds it to the final set of results if its p-value is smaller than the significance threshold $L[k]$ (Lines 10–11). This procedure is repeated until the size of the results reaches n. Even, if it scans candidates in decreasing order of their coverage, COVER_G controls the false discovery rate at level α by adjusting their p-values using the Benjamini-Yekutieli procedure.

We highlight that COVER_G has a $(1 - 1/e)$ approximation guarantee [1] as our problem has a one-to-one reduction to Maximum Coverage (proof sketch in Sect. 3.4).

The worst-case complexity of COVER_G is $O(m \cdot p + m \cdot log\ m + m \cdot n)$. The complexity of calculating a single p-value depends on the type of test and the size of the compared groups. Assuming the worst case complexity for computing a single p-value is p, the term $m \cdot p$ denotes the complexity of calculating the p-values of all m candidates. The term $m \cdot log\ m$ is for sorting the candidates by their p-values and the term $m \cdot n$ is the complexity of the greedy scans to select the next best candidate at each step.

Algorithm 2: Greedy coverage algorithm (COVER_G) – illustrated with the Benjamini-Yekutieli correction

---

Input: a Request R, a dataset D, a significance level α, number of desired results n

Output: C

1 $allCandidates \leftarrow GenerateCandidates(R, D)$
2 $Candidates \leftarrow ComputePvalues(allCandidates, \alpha)$
3 $L = Sortbypval\ (Candidates)$
4 $C \leftarrow \emptyset$
5 $m \leftarrow |Candidates|$

6 $k = \text{argmax}_{0 \leq j \leq m}\ L[j] \leq \frac{\alpha \times j}{m} \left(\sum_{i=1}^{m} 1/i \right)^{-1}$

7 **while** $|C| \leq n$ **do**
8       $c^* = \text{argmax}_{c \in Candidates}\ |cover(C.groups \cup \{c.groups\}, D)|$
9       $Candidates \leftarrow Candidates \setminus \{c^*\}$
10   **if** $c^*.pval \leq L[k]$ **then**
11          $C \leftarrow C \cup \{c^*\}$

12 **return** C

4.3   Algorithm COVER_α

To solve Problem 3, we revisit α-investing, a method for multiple hypothesis testing that controls specifically $mFDR$. α-investing was originally introduced by Foster and Stine [11], and generalized by Zhao et al. for data exploration [32]. Intuitively, α-investing works as follows: it starts with an initial wealth, set to $\eta.\alpha$, then at each step j a specific threshold α_j is defined, which is below the current available wealth. If the null hypothesis is accepted $(p_j > \alpha_j)$ a fraction of the invested value is lost and is subtracted from the current available wealth $W(j)$. If the null hypothesis is rejected $(p_j \leq \alpha_j)$, we obtain a "return" on investment $\omega \leq \alpha$. The testing procedure stops when the available α-wealth is totally consumed, i.e., reaches 0. We choose $\eta = 1$ and $\omega = \alpha$ as it was shown in [11], any α-investing algorithm controls $mFRD_\eta$ at level α for $W(0) = \eta \cdot \alpha$ and $\omega = \alpha$ for any $\eta, \alpha \in [0, 1]$.

The key idea of COVER_α is to re-adjust the quantities of the α-wealth that are invested according to *the coverage* of each selected candidate. Different policies for investing the wealth could be explored [32]. In our solution, we design COVER_α in such a way that it invests more α-wealth on candidates that bring higher coverage of the input data D. In Sect. 5, we also implement existing variants and compare them to COVER_α.

Algorithm 3 contains the pseudo-code of COVER_α. It starts with generating the set of candidates by calling the sub-routine *GenerateCandidates()* (Line 1). However, unlike COVER_G, it does not compute p-values of all candidates as it relies on $mFDR$. It initializes α-wealth (Line 2) with η set to 1. The adjustment value of the hypothesis testing is initialized with a fixed parameter λ (Line 3) which controls how much of the available α-wealth is invested during each

---

Algorithm 3: α-investing coverage algorithm (COVER_α)

Input: a request R, a dataset D, a significance level α, number of results n,
      parameters λ and η

Output: C

1 $Candidates \leftarrow GenerateCandidates(R, D)$

2 $W(0) \leftarrow \eta \cdot \alpha$

3 $\alpha^* = \frac{W(0)}{\lambda + W(0)}$

4 $C \leftarrow \emptyset$

5 $j \leftarrow 1$

6 **while** $W(j-1) > 0$ *and* $|C| \leq n$ **do**

7     $c^* = \text{argmax}_{c \in Candidates} |cover(C.groups \cup \{c.groups\}, D)|$

8     $Candidates \leftarrow Candidates \setminus \{c^*\}$

9     $\alpha_j = \alpha^* (\frac{|cover(c^*.groups, D)|}{|D|})^{1/2}$

10   **if** $W(j-1) - \frac{\alpha_j}{1-\alpha_j} \geq 0$ **then**

11   **if** $c^*.pval \leq \alpha_j$ **then**

12          $W(j) \leftarrow W(j-1) + \alpha$

13          $C \leftarrow C \cup \{c^*\}$

14   **else**

15          $W(j) \leftarrow W(j-1) - \frac{\alpha_j}{1-\alpha_j}$

16    $j \leftarrow j + 1$

17 **return** C

---

step. In our experiments (Sect. 5.4), we vary the parameter λ and show that higher values are preferred. The set of returned results is initialized with the empty set (Line 4). While the number of results added to the set is less than the number of desired results and the value of α-wealth remains positive, the algorithm picks the next candidate that maximizes coverage (Line 7) and removes it from the set of candidates (Line 8). It then sets the current α_j value according to coverage of the selected candidate (Line 9) and checks the availability of α-wealth (Line 10). Unlike COVER_G, the threshold for each hypothesis test in not fixed but is dependent on how much coverage the candidate adds to the current solution. COVER_α compares the p-value of the selected candidate c^* with its threshold α_j. If the null hypothesis is rejected (Line 11), α is added to the current α-wealth (Line 12), and c^* is added to the set of results (Line 13). Otherwise, if the null hypothesis is accepted, the amount $\frac{\alpha_j}{1-\alpha_j}$ is lost and is subtracted from the α-wealth accordingly (Line 15).

COVER_α greedily scans n times the set of candidates of size m, retrieves at each step the best candidate c^* that has the maximum coverage and computes its p-value with an $O(p)$ worst-case complexity. This gives us a $O(m \cdot n \cdot p)$ worst-time complexity for COVER_α which is much faster than COVER_G. It benefits from computing p-values on-the-fly and only for candidates with the highest coverage at each iteration.

5 Experiments

Our experiments aim to: (1) demonstrate the expressivity of GROUPTEST on realistic scenarios (Sect. 5.1); (2) study the results of VAL_C in terms of coverage and scalability and compare it to different baselines (Sect. 5.3); (3) study hypothesis significance and the interplay between coverage and significance for COVMAX algorithms (Sect. 5.4).

Most of our experiments use request R_8 in Table 3 that returns the highest number of results. Our code and complete results are available on our Github repository.

5.1 Addressing Information Needs

We describe two scenarios using a Traditional multiple hypothesis algorithm based on Benjamini-Yekutieli (TRAD_BY) and COVER_G to illustrate the expressivity of GROUPTEST.

Group Evolution. We use TRAD_BY with a multiple mean F-test to verify the request "Groups whose average rating for a movie genre changes monthly (in a 3-month period)" (akin to R_4 in Table 3 with an unlimited n). Table 4 reports for each movie genre the number of groups that exhibit the monthly change along with examples and the movie genres for which that change is observed. For instance, the average rating of [35–44] aged females who rated Comedies from the 60's changes monthly.

Table 4. Number and examples of groups per genre that satisfy the request "Groups whose average rating for a movie genre changes monthly (in a 3-month period)".

Genres	#Groups	Example groups
All	6	[18–24] aged females who rated 50's movies
Drama	2	[35–44] aged females who rated 70's movies
Horror	2	Males who rated 80's movies
Action	2	Reviewers whose occupation is in customer-service
Science-Fiction	2	Male students under 18
Comedy	4	[35–44] aged females who rated 60's movies

Group Comparison. We use COVER_G with a two-sample Kolmogorov-Smirnov test to compare rating distributions of group pairs within the same age category (akin to R_8 in Table 3 with $n = 20$). Table 5 reports for each age category, the number of returned pairs along with examples. For instance, we find 14 pairs for age [25–34] among which the two groups who rated War movies and 90's Thrillers significantly differ in their rating distribution.

Table 5. Number and examples of pairs per age that satisfy the request "Group pairs whose rating distribution for Drama movies differs in the same season (Summer)".

Age	#Pairs	Example pairs
<18	2	(Drama movies - Comedy movies)
18–24	7	(Male college student- Male users)
25–34	14	(War movies - 90's Thriller movies)
35–44	10	(Males who rated 80's movies - Male who rated 70's movies)
45–49	5	(Males who rated 2000's movies - Females who rated 90's movies)
50–55	4	(Users who rated 90's Thrillers - Users who rated 50's movies)

5.2 Experimental Setup

Datasets. We report results on MovieLens'1M. Similar results were found on Yelp, TAFENG, and BookCrossing datasets. Our complete results are available on our Github repository. The MovieLens'1M dataset contains user-movie ratings collected from a movie recommendation service. It contains around $1M$ ratings given by $6,040$ users to $3,900$ movies. The data also contain user attributes: gender, age group, occupation and location, as well as item attributes which correspond to a set of genres.

Evaluation Measures. We examine for each request, the groups that are found in terms of (1) their significance (min/max/sum p-values), (2) data coverage, (3) power (#true positives/#results in ground-truth) and FDR (#false positives/#results), (4) response time. All results are averages of 10 runs.

VALMIN Algorithm Variants. In Sect. 5.3 we implement two variants of VAL_C, one with Bonferroni noted VAL_C_BN and the other one with Benjamini-Yekutieli noted VAL_C_BY. We compare the algorithm against an adaptation of the Subfamily wise Multiple Testing Procedure (SMT) [29] where we add a constraint on coverage cov_{min} (called SMT$_{cov}$). This method controls the FWER in multiple testing where hypotheses are organized into families. One null hypothesis is to be rejected in each family. The procedure has two steps:

– Step 1: finds r^* as

$$r^* = \operatorname*{argmax}_{r} (\sum_{i=1}^{r} p_i^{min} \cdot |F_i| \leq \alpha) \qquad (8)$$

where r is the number of families, F_i is the ith family and p_i^{min} is the smallest $p-value$ in that family. r^* is the perfect number of families.
– Step 2: Reject $h_1^{min}, ..., h_{r^*}^{min}$, where h_i^{min} is the rejected hypothesis of family F_i.

In our adaptation, the families are generated randomly using one hyper parameter. It represents the number of hypotheses in each family. We compare different variants by varying the hyper parameter in $\{10, 50, 100, 500, 1000, 5000\}$. In the experiments, we compare VAL_C against the traditional correction method, referred to as TRAD with Benjamini-Yekutieli (TRAD_BY), the original SMT and its adaptation (SMT$_{cov}$) varying cov_{min} in $\{0.1, 0.3, 0.5, 0.7, 0.9\}$.

CovMax Algorithm Variants. In Sect. 5.4 we compare variants of COVER_α by varying the hyper parameter λ in $\{20, 50, 100, 200, 500\}$. We compare our algorithms, the best COVER_α and the two variants of COVER_G, one with Bonferroni noted COVER_G_BN and the other one with Benjamini-Yekutieli noted COVER_G_BY, against the traditional correction method, referred to as TRAD. Similarly to COVER_G, we implemented two variants of TRAD, one with Bonferroni correction that we note TRAD_BN and the other with Benjamini-Yekutieli that we note TRAD_BY. We also implemented previously proposed α-investing policies [32] as baselines for our comparisons:

- β-Farsighted: it ensures that at each step a fraction β of the available α-wealth is preserved for future tests.
- γ-Fixed: it assigns a fixed budget defined as a function of γ for each performed test.
- δ-Hopeful: it assigns the whole current available α-wealth to each hypothesis with the *hope* that at least one of the next δ null hypotheses is rejected.
- ϵ-Hybrid: it adjusts the budget assigned to the tests based on an estimated data randomness and chooses between γ-Fixed and δ-Hopeful using a randomness threshold ϵ.
- ψ-Support: it adjusts the budget of each hypothesis based on its support population, i.e., the number of data points that are used to perform the test.

Each α-investing policy was tested with different parameter values (in Table 6). We selected the best value (last column in the table) based on power and FDR. For instance, for our algorithm COVER_α, we observed that $\lambda = 500$ yields the best power and FDR since smaller λ make COVER_α consume its α-wealth faster. α is set to 0.05.

Setup. We use a x86_64 GNU/Linux Debian server with an Intel(R) Xeon(R) Gold 6130 CPU @ 2.10 GHz, and 394 GB memory.

5.3 VALMIN Results

Study of SMT$_{cov}$ Variants. In this section, we study the significance and coverage of the different SMT$_{cov}$ variants by varying the value of the size of families (number of hypotheses in each family).

As in [32], we use TRAD with Bonferroni (TRAD_BN) as ground-truth since BN is the most conservative correction method and compare the results of

Table 6. Parameters for α-investing policies.

Investigation policies	Parameter	Values	Best Value
COVER_α	λ	$20, 50, 100, 200, 300, 500$	**500**
β-Farsighted	β	$0.25, 0.5, 0.75, 0.9$	**0.9**
γ-Fixed	γ	$20, 50, 100, 200, 300, 500$	**500**
δ-Hopeful	δ	$20, 50, 100, 200, 300, 500$	**500**
ϵ-Hybrid	ϵ	$0.25, 0.5, 0.75, 0.9$	**0.75**
ψ-Support	ψ	$1/2, 1/3, 1/4, 1/5, 1/6$	**1/2**

SMT$_{cov}$ variants. We also consider data samples ranging from 10% to 100% of *Candidates* and report the results for R$_8$.

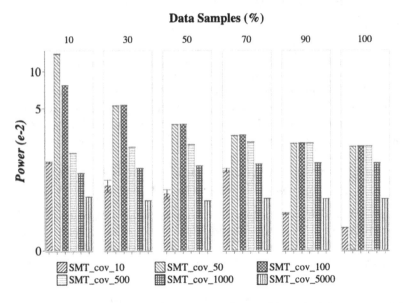

Fig. 3. Impact of coverage optimization on significance (Power) with $cov_{min} = 0.5$ and number of results $n = 100$ for different percentages of data samples on `MovieLens'1M` for R$_8$.

Figure 3 reports power for $n = 100$. FDR average results are all the same and equal to zero. The main observation is that SMT$_{cov}$ with family size of 50 performs better than all other variants. Significance power decreases as the value of family size increases. Large values $(500, 1000, 5000)$ make the families too massive with hypotheses. As SMT$_{cov}$ rejects one and only one hypothesis in each family, many valid and potential true discoveries will be skipped. This will result in a weak power score. Similar effect can be observed with small values of size (10). The families will be, in this case, narrow but many true discoveries may be part of the same family.

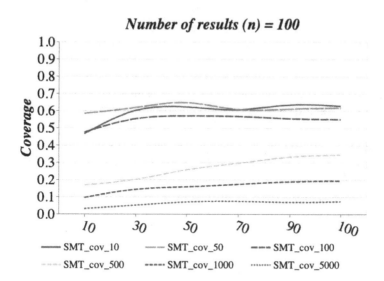

Fig. 4. Coverage as a function of the number of data samples with $cov_{min} = 0.5$ and number of results $n = 100$ on MovieLens'1M for R_8.

Figure 4 reports results of coverage as a function of number of data samples with results number $n = 100$. We note that the coverage is nearly constant as the number of samples increases for all cases. One can observe that SMT_{cov} variants with small family size achieve a coverage greater, and in some cases slightly smaller, than $cov_{min} = 0.5$ while variants with large size don't cover more than 35% of data.

VAL_C Hypothesis Significance. We study, in this part, the significance by comparing between VAL_C_BY, the best variant of SMT and SMT_{cov} (with family size 50) and TRAD_BY.

Figures 5, 6 report power and FDR for $n = 20$ (**left**) and $n = 100$ (**right**) with $cov_{min} = 0.7$ respectively. We observe that, for small number of results, SMT and its adaptation SMT_{cov} outperform VAL_C_BY in terms of power and FDR. One can also observe that SMT behaves similarly to TRAD_BY in terms of power. On the other hand, for larger results values, the performances of VAL_C_BY are better and are slightly outperformed by the baselines for power but still much worse for FDR regardless of the sample size.

One possible explanation of this difference is the closeness between the ground-truth and the results of the baseline SMT_{cov}. Indeed, as explained in Sect. 5.2 SMT procedure controls the Family-Wise Error rate (FWER) as well as TRAD_BN that is used to generate the ground-truth. In contrary, VAL_C_BY

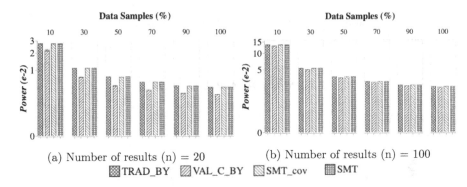

Fig. 5. Impact of coverage optimization on significance (Power) with $cov_{min} = 0.7$, number of results $n = 20$ **(left)** and $n = 100$ **(right)** for different percentages of data samples on MovieLens'1M for R_8.

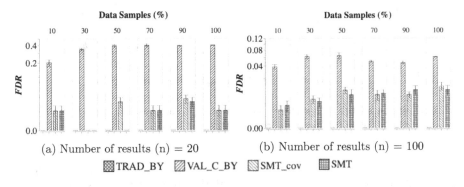

Fig. 6. Impact of coverage optimization on significance (FDR) with $cov_{min} = 0.7$, number of results $n = 20$ **(left)** and $n = 100$ **(right)** for different percentages of data samples on MovieLens'1M for R_8.

controls FDR which makes it disadvantageous in terms of significance. For small number of results, SMT_{cov} behaves approximately like TRAD_BN due to the similarity of their policies while VAL_C_BY explores more the set of potential discoveries in order to satisfy the coverage constraint making by that more false discoveries. As n increases, SMT_{cov} explores more the set of candidates resulting in a relatively increase of its false discoveries causing by that a more similar power rate comparing to VAL_C_BY.

VAL_C Data Coverage. Figure 7 reports results of coverage as a function of number of data samples with results number $n = 20$ **(left)** and $n = 100$ **(right)**.

We observe that for small number of results the coverage is nearly constant for SMT_{cov} and increases for VAL_C_BY as the number of samples increases. One can see that VAL_C_BY achieves a coverage rate greater than the threshold $cov_{min} = 0.7$ while the best SMT_{cov} algorithm doesn't satisfy this constraint.

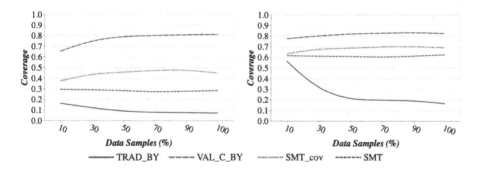

Fig. 7. Coverage as a function of the number of data samples with $cov_{min} = 0.7$, number of results $n = 20$ **(left)** and $n = 100$ **(right)** on MovieLens'1M for R_8.

Obviously, one can see that TRAD_BY and SMT are the worst in terms of coverage as they are not developed to satisfy it. For large number of results, VAL_C_BY still outperform the baseline and satisfies the coverage constraint for all data samples unlike SMT_{cov} which does not satisfy it for small and medium data samples $(10\%, 30\%, 50\%)$. One possible explanation is that SMT_{cov} returns more granular and by that more overlapping groups. Indeed, significant groups with high coverage rate may be part of families where more significant groups but less covered are also members. As the most significant group is chosen from each family, the granular one is finally returned.

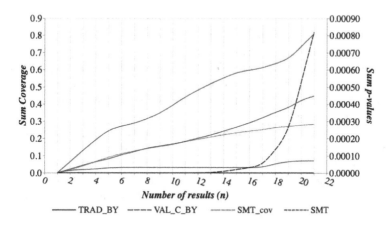

Fig. 8. Coverage and p-values as a function of the number of results n with $cov_{min} = 0.7$, data samples $= 100\%$ on MovieLens'1M for R_8.

We now examine simultaneously the evolution of the cumulative coverage and p-values. Results are depicted in Fig. 8 for $n = 20$. The graph clearly shows that, VAL_C_BY reaches cov_{min} by iteration 19 while SMT_{cov} never does.

One can also observe that after satisfying the constraint, the cumulative p-value of VAL_C_BY increases immediately while the baselines' still equal to zero. One possible reason for that is that larger groups may be less significant than smaller ones even if they control the multiple testing error. We performed other experiments on greater values of n and we observed the same results for cumulative p-values. We also observed that both SMT_{cov} and VAL_C_BY reach the coverage threshold at the same time.

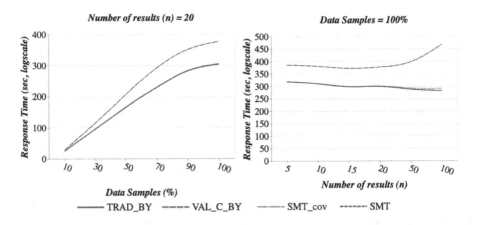

Fig. 9. Response time as a function of number of data samples (**left**) and number of results n (**right**) on MovieLens'1M for R_8 with $cov_{min} = 0.7$.

VAL_C Scalability. We finally study the evolution of response time as a function of input data sample (Fig. 9 - left) and of number of results n (Fig. 9 - right). First it shows that response time increases with the increase of data sample but remains mostly constant with the increase of n. It also shows that SMT_{cov} outperforms VAL_C_BY in terms of scalability regardless of the data sample or returned results.

In these experiments, we only report results with $cov_{min} = 0.7$. We varied the value of cov_{min} in $\{0.1, 0.3, 0.5, 0.9\}$. The results are mostly similar than the reported ones.

In summary, the use of VAL_C_BY to solve VALMIN is more appropriate than SMT_{cov}. Indeed, even if it has smaller power significance for small values of n, it succeeds to achieve the coverage constraint while SMT_{cov} fails to do so.

5.4   CovMax Results

Study of COVER_α variants. In this section, we study the significance, coverage and scalability of the different COVER_α variants by varying the value of the hyper parameter λ. We set TRAD with Bonferroni (TRAD_BN) as a

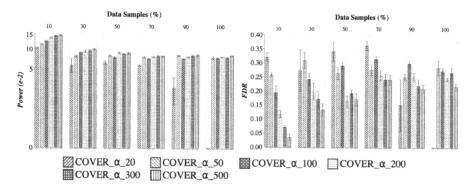

Fig. 10. Impact of coverage optimization on significance (Power and FDR) with results number $n = 50$ for different percentages of data samples on MovieLens'1M for R_8.

ground-truth, and compare the results of COVER$\_\alpha$ variants. We consider data samples ranging from 10% to 100% of *Candidates* and report the results for R_8.

Figure 10 reports power and average FDR. The main observation is that COVER$\_\alpha_{500}$ performs better than all other COVER$\_\alpha$ variants with a different λ parameter on both average FDR and power. Small values of λ makes COVER$\_\alpha$ consume its α-wealth quickly and thus makes its discovery power smaller and its average FDR more important.

We now examine simultaneously the evolution of the cumulative coverage and p-values. Results are depicted in Fig. 11. The graph clearly shows that all COVER$\_\alpha$ variants perform similarly and reach full coverage by iteration 49. We also notice that the difference of coverage between the variants is significant in the first iterations but it narrows to the point of being negligible for high iterations. Indeed at the iteration 50, COVER$\_\alpha\_50$ covers 99% of data while COVER$\_\alpha\_500$ covers 97%. One can also observe that in terms of sum of p-values the difference between variants is more significant and COVER$\_\alpha\_500$ outperforms all the other variants.

The last experiment studies evolution of response time as a function of input data samples (Fig. 12 - left) and of number of results n (Fig. 12 - right). It shows that COVER$\_\alpha_{20}$ is the worst performer as it consumes a large amount of its α-wealth in earlier iterations. It is left with a very small wealth that requires many iterations to reach n results whose p-values quality. The figure shows that the remaining COVER$\_\alpha$ variants have practically the same response time with a slight advantage for COVER$\_\alpha_{50}$ and COVER$\_\alpha_{100}$.

This final result shows that an α-investing strategy with a high λ value can attain high coverage while ensuring sound group testing in reasonable times.

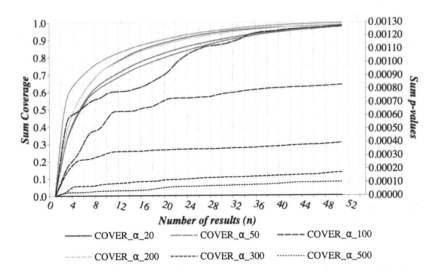

Fig. 11. Coverage and p-values as a function of the number of results n, data samples $= 100\%$, number of results $n = 50$ on `MovieLens`'1M for R_8.

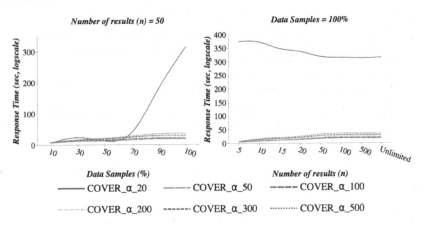

Fig. 12. Response time as a function of number of data samples (**left**) and number of results n (**right**) on `MovieLens`'1M for R_8.

COVER_α Hypothesis Significance. We study the impact of adjustment and coverage on significance.

We first study the impact of adjustment on significance using the traditional corrections (TRAD and COVER_G). We ran all requests in Table 3. The complete results are shown in Table 7.

The first observation is that in all two-sample tests, the p-value computation protocol (Sect. 3.3) reduces the number of candidates by one order of magnitude. We also observed that both TRAD variants return by far the highest number of results when n is unlimited, since they do not set a coverage constraint. Since TRAD_BY is less stringent, it consistently yields more results than TRAD_BN. This is apparent in the sum of p-values that are significantly higher for TRAD_BY. In some cases (R_8), the smallest p-value returned by COVER_G is higher than TRAD since it optimizes coverage and does not necessarily reach the lowest p-values. The second observation is that when n is capped, COVER_G achieves higher coverage than both TRAD variants (up to 5x), which leads us to conclude that combining coverage maximization with significance adjustment is necessary for sound group testing.

Table 7. Results of running all requests in Table 3 on `MovieLens'1M` (equal values are shown only once in each cell)

Ri	n	Methods	#allCandidates #Candidates	Benjamini-Yekutieli (BY)/Bonferroni (BN)				
				#Results no-adjustment/BY/BN	Min p-value	Max p-value	Sum p-value	Total Cov
R_1	Unlimited	TRAD	1663	907/583/337	4.18 e-35	4.23 e-03/ 5.5 e-05	0.3/1.91 e-03	1
		COVER_G	1663	3	4.78 e-25	1.9 e-13	1.9 e-13	1
	20	TRAD		20	4.18 e-35	4.43 e-22	6 e-22	0.98
		COVER_G		3	4.78 e-25	1.9 e-13	1.9 e-13	1
R_2	Unlimited	TRAD	1 329 259	20 505/8208/1351	7.28 e-22	1.92 e-03/2 e-06	3.62/6.57 e-04	0.99/0.94
		COVER_G	108 478	17/23	1.14 e-17/3.93 e-04	2 e-06/3.02 e-07	8.88 e-04/4 e-06	0.99/0.94
	15	TRAD		15	7.28 e-22	4.08 e-17	1.34 e-16	0.22
		COVER_G		15	1.14 e-17	3.93 e-04/2 e-06	8.88 e-04/4 e-06	0.99/0.92
R_3	Unlimited	TRAD	198	6/4/2	2 e-03	9.09 e-03/4.01 e-03	2.22 e-02/4.01 e-03	0.02
		COVER_G	17	2/1	2 e-03	1.8 e-02/2 e-03	2 e-02/2 e-03	0.12/0.02
R_4	Unlimited	TRAD	749 749	19 472/5320/616	1.95 e-15	1.26 e-03/2.45 e-06	1.80/4.45 e-04	1/0.96
		COVER_G	44 368	7	1.95 e-15	2.93 e-05/2.03 e-06	6.93 e-05/3.75 e-06	1/0.96
	5	TRAD		5	1.95 e-15	1.95 e-15	9.77 e-15	0.17
		COVER_G		5	1.95 e-15	3.93 e-05/2.02 e-06	4.78 e-05/2.66 e-06	0.97/0.92
R_5	Unlimited	TRAD	6 344	79/70/39	1.67 e-18	8.39 e-03/4.01 e-04	0.11/2.14 e-03	0.05/0.04
		COVER_G	6 344	25/15	1.67 e-18	8.39 e-3/4.01 e-01	3.71 e-02/8.84 e-04	0.05/0.04
	10	TRAD		10	1.67 e-18	1.58 e-11	5.14 e-11	0.01
		COVER_G		10	5.3 e-16/1.67 e-18	3.54 e-03/1.27 e-04	5.16 e-03/1.76 e-04	0.04/0.03
R_6	Unlimited	TRAD	429 025	6 076/1412/1008	0	1.22 e-03/5 e-06	0.2/4.3 e-05	0.85/0.74
		COVER_G	42 657	43/36	0	8.09 e-04/5 e-06	6.25 e-03/1.1 e-05	0.85/0.74
	20	TRAD		20	0	0	0	0.33
		COVER_G		20	0	8.09 e-04/5 e-06	2.85 e-03/5 e-06	0.76/0.66
R_7	Unlimited	TRAD	85 908	85 908	0	3.37 e-13	5.14 e-11	1
		COVER_G	85 908	1	0	0	0	1
R_8	Unlimited	TRAD	695 772	26 772/16 397/4 790	0	2.83 e-03/1.86 -06	6.58/1.3 e-03	1/0.97
		COVER_G	49 260	33/51	6.21 e-11/3.34 e-18	2.44 e-03/1.59 e-06	6.4 e-03/1.11 e-05	1/0.97
	10	TRAD		10	0	2.02 e-22	5.8 e-22	0.21
		COVER_G		10	6.21 e-11/6.83 e-14	2.44 e-03/1.59 e-06	2.79 e-03/3.19 e-06	0.87 / 0.72

We now examine how optimizing coverage affects significance. We compare the results of COVER_α, COVER_G and α_investing policies. Figure 13 reports power and FDR. We observe that COVER_α performs better than all other α-investing variants on both power and FDR. The second observation is that COVER_α outperforms COVER_G_BY regardless of the sample size. Additionally, COVER_α attains similar power as COVER_G_BN, especially for smaller sample sizes. To better understand the difference between groups returned by TRAD_BY, COVER_G and COVER_α, we analyze the occurrence of attribute-value pairs appearing in the results of R_8. For instance for

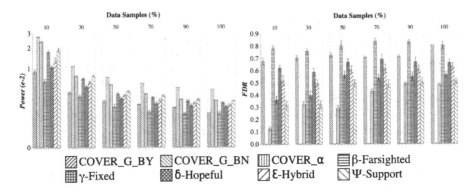

Fig. 13. Impact of coverage optimization on significance (Power and FDR) with results number $n = 20$ for different percentages of data samples on MovieLens'1M for R_8.

$n = 20$, we find that TRAD_BY returns groups that cover a total of 11 attribute-value pairs such as gender (male, female), occupation (doctor, scientist, lawyer), age ([25–34], [45–49], [35–44]) and movies of 2 decades (90's, 70's). On the other hand, COVER_G_BY and COVER_α return groups that cover 24 and 23 attribute-value pairs respectively. Groups contain all gender and age values, more user occupation attributes (6 for COVER_G_BY and 5 for COVER_α) and all movies from the 50's to the 2000's.

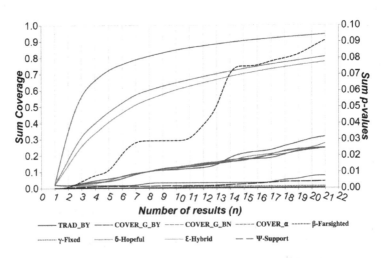

Fig. 14. Coverage and p-values as a function of the number of results n on MovieLens'1M for R_8.

COVER_α Data Coverage. We now seek to find if our formulation of Cov-Max hurts the significance of retrieved groups. We compare the results of R_8 using TRAD_BY, COVER_G, COVER_α and all other α-investing policies.

We examine simultaneously the evolution of the cumulative coverage and p-values. Results are depicted in Fig. 14. The graph clearly shows that COVER_α performs closely to COVER_G_BN and reaches 79% coverage by iteration 20. We see that COVER_G_BY is slightly better as it achieves 92% coverage. Additionally, one can observe that TRAD_BY and all α-policies cover, at best, only 9% and 30% of the input data respectively. We also notice that β-Farsighted invests most of the α-wealth in testing insignificant results and that COVER_G_BY is the second worst.

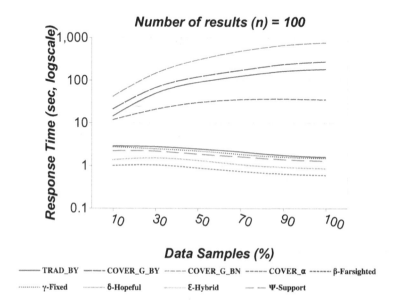

Fig. 15. Response time as a function of number of data samples on MovieLens'1M for R_8.

COVER_α Scalability Analysis. Our last experiment studies evolution of response time as a function of the number of data samples (Fig. 15) and the number of results n (Fig. 16). The first observation is that the response time of COVER_α remains stable with the increase of both the sample size and the number of results n. Indeed, COVER_α computes p-values only for candidates with the highest coverage. Therefore, it clearly outperforms COVER_G_BN, COVER_G_BY and TRAD_BY. However, COVER_α performs worse than the other α-investing algorithms, which is mainly due to COVER_α performing at each step a scan over all the remaining candidates to select the one with the highest coverage. But despite, being slightly slower, COVER_α performs marginally better in terms of (1) coverage: reaches almost twice the coverage of the best performing α-investing (2) FDR: makes up to 8 times less false discoveries than β-Farsighted and up to 4 times less than ϵ-Hybrid for instance.

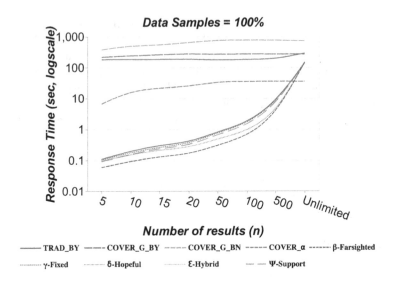

Fig. 16. Response time as a function of number of results n on MovieLens'1M for R_8.

In summary, our results confirm that COVER_α is the method of choice to attain high coverage of the dataset while ensuring sound group testing in reasonable times.

6   Conclusion and Future Work

We developed GROUPTEST, a framework that enables data-driven discoveries by combining statistical testing with optimizing data coverage. GROUPTEST accommodates different types of tests and group aggregations and comparisons.

We believe our work to be the first to propose a generic and principled framework to multiple hypothesis testing on large datasets. Our framework opens several directions. An immediate extension is to understand the relationship between α-investing and top-n. We are formalizing the problems of finding the best n, i.e., the one that is most likely to maximize significance and coverage. One other question is to integrate our information needs described in Sect. 5.1 as first-class citizens in our model. This would allow us to address requests such as "Female groups whose monthly average rating for Comedies remains stable over a given year", or "Males and Females who exhibit the same variance for Drama movies in a 10-year period". These requests will require to carefully model time and account for it in hypothesis significance and data coverage.

References

1. Ageev, A.A., Sviridenko, M.I.: Approximation algorithms for maximum coverage and max cut with given sizes of parts. In: Cornuéjols, G., Burkard, R.E., Woeginger, G.J. (eds.) IPCO 1999. LNCS, vol. 1610, pp. 17–30. Springer, Heidelberg (1999). https://doi.org/10.1007/3-540-48777-8_2
2. Agrawal, R., Gehrke, J., Gunopulos, D., Raghavan, P.: Automatic subspace clustering of high dimensional data for data mining applications, vol. 27. ACM (1998)
3. Amer-Yahia, S., Kleisarchaki, S., Kolloju, N.K., Lakshmanan, L.V., Zamar, R.H.: Exploring rated datasets with rating maps. In: Proceedings of the 26th International Conference on World Wide Web, pp. 1411–1419. International World Wide Web Conferences Steering Committee (2017)
4. Beliakov, G., James, S., Mordelová, J., Rückschlossová, T., Yager, R.R.: Generalized bonferroni mean operators in multi-criteria aggregation. Fuzzy Sets Syst. **161**(17), 2227–2242 (2010)
5. Benjamini, Y., Yekutieli, D.: The control of the false discovery rate in multiple testing under dependency. Ann. Stat. **29**, 1165–1188 (2001)
6. Boley, M., Mampaey, M., Kang, B., Tokmakov, P., Wrobel, S.: One click mining: interactive local pattern discovery through implicit preference and performance learning. In: Proceedings of the ACM SIGKDD Workshop on Interactive Data Exploration and Analytics, pp. 27–35. ACM (2013)
7. Bron, C., Kerbosch, J.: Algorithm 457: finding all cliques of an undirected graph. Commun. ACM **16**(9), 575–577 (1973)
8. Chekuri, C., Quanrud, K., Zhang, Z.: On approximating partial set cover and generalizations. arXiv preprint arXiv:1907.04413 (2019)
9. Colquhoun, D.: An investigation of the false discovery rate and the misinterpretation of p-values. R. Soc. Open Sci. **1**(3), 140216 (2014)
10. Di Leo, G., Sardanelli, F.: Statistical significance: p value, 0.05 threshold, and applications to radiomics-reasons for a conservative approach. Eur. Radiol. Exp. **4**(1), 1–8 (2020)
11. Foster, D., Stine, R.A.: Alpha-investing: a procedure for sequential control of expected false discoveries. J. Roy. Stat. Soc. Ser. B Stat. Methodol. **70**(2), 429–444 (2008)
12. Goyal, A., Bonchi, F., Lakshmanan, L.V.: Discovering leaders from community actions. In: Proceedings of the 17th ACM Conference on Information and Knowledge Management, pp. 499–508. ACM (2008)
13. Greenland, S., et al.: Statistical tests, P values, confidence intervals, and power: a guide to misinterpretations. Eur. J. Epidemiol. **31**(4), 337–350 (2016). https://doi.org/10.1007/s10654-016-0149-3
14. Hämäläinen, W., Webb, G.I.: A tutorial on statistically sound pattern discovery. Data Min. Knowl. Disc. **33**(2), 325–377 (2018). https://doi.org/10.1007/s10618-018-0590-x
15. Hochbaum, D.S., Pathria, A.: Analysis of the greedy approach in problems of maximum k-coverage. Nav. Res. Logist. (NRL) **45**(6), 615–627 (1998)
16. Jafari, M., Ansari-Pour, N.: Why, when and how to adjust your p values? Cell J. (Yakhteh) **20**(4), 604 (2019)
17. Jiang, D., et al.: Cohort query processing. Proce. VLDB Endow. **10**(1), 1–12 (2016)
18. Kamat, N., Jayachandran, P., Tunga, K., Nandi, A.: Distributed and interactive cube exploration. In: 2014 IEEE 30th International Conference on Data Engineering (ICDE), pp. 472–483. IEEE (2014)

19. Karp, R.M.: Reducibility among combinatorial problems. In: Miller, R.E., Thatcher, J.W., Bohlinger, J.D. (eds.) Complexity of Computer Computations, pp. 85–103. Springer, Boston (1972). https://doi.org/10.1007/978-1-4684-2001-2_9

20. Meijer, R.J., Goeman, J.J.: Multiple testing of gene sets from gene ontology: possibilities and pitfalls. Briefings Bioinform. **17**(5), 808–818 (2016)

21. Mieth, B., et al.: Combining multiple hypothesis testing with machine learning increases the statistical power of genome-wide association studies. Sci. Rep. **6**(1), 1–14 (2016)

22. Newman, M.E.J.: Detecting community structure in networks. Eur. Phys. J. B **38**(2), 321–330 (2004). https://doi.org/10.1140/epjb/e2004-00124-y

23. Nikolaev, A.G., Gore, S., Govindaraju, V.: Engagement capacity and engaging team formation for reach maximization of online social media platforms. In: KDD, pp. 225–234 (2016)

24. Pedreira, P., Croswhite, C., Bona, L.: Cubrick: indexing millions of records per second for interactive analytics. Proc. VLDB Endow. **9**(13), 1305–1316 (2016)

25. Pellegrina, L., Riondato, M., Vandin, F.: Hypothesis testing and statistically-sound pattern mining (tutorial). In: Proceedings of the 25th ACM SIGKDD International Conference on Knowledge Discovery & Data Mining, KDD 2019, Anchorage, AK, USA, 4–8 August 2019, pp. 3215–3216 (2019)

26. Roquain, E.: Type i error rate control for testing many hypotheses: a survey with proofs. Journal de la Société Française de Statistique **152**(2), 3–38 (2011)

27. Srikant, R., Agrawal, R.: Mining generalized association rules. Futur. Gener. Comput. Syst. **13**(2–3), 161–180 (1997)

28. Webb, G.I.: Discovering significant patterns. Mach. Learn. **68**(1), 1–33 (2007)

29. Webb, G.I., Petitjean, F.: A multiple test correction for streams and cascades of statistical hypothesis tests. In: Proceedings of the 22nd ACM SIGKDD Conference on Knowledge Discovery and Data Mining, San Francisco, USA, August 2016, pp. 1255–1264 (2016)

30. Xin, D., Shen, X., Mei, Q., Han, J.: Discovering interesting patterns through user's interactive feedback. In: Proceedings of the 12th ACM SIGKDD International Conference on Knowledge Discovery and Data Mining, pp. 773–778. ACM (2006)

31. Zgraggen, E., Zhao, Z., Zeleznik, R., Kraska, T.: Investigating the effect of the multiple comparisons problem in visual analysis. In: Proceedings of the 2018 CHI Conference on Human Factors in Computing Systems, pp. 1–12 (2018)

32. Zhao, Z., Stefani, L.D., Zgraggen, E., Binnig, C., Upfal, E., Kraska, T.: Controlling false discoveries during interactive data exploration. In: Proceedings of the 2017 ACM International Conference on Management of Data, SIGMOD Conference 2017, Chicago, IL, USA, 14–19 May 2017, pp. 527–540. ACM (2017)

Efficiently Identifying Disguised Missing Values in Heterogeneous, Text-Rich Data

Théo Bouganim[1,2(✉)], Helena Galhardas[3], and Ioana Manolescu[1,2]

[1] Inria Saclay Ile de France, Palaiseau 91120, France
{theo.bouganim,ioana.manolescu}@inria.fr
[2] Institut Polytechnique de Paris (IPP), Paris, France
[3] INESC-ID & IST, Universidade de Lisboa, Lisbon, Portugal
hig@inesc-id.pt

Abstract. Digital data is produced in many data models, ranging from highly structured (typically relational) to semi-structured models (XML, JSON) to various graph formats (RDF, property graphs) or text. Most real-world datasets contain a certain amount of *null* values, denoting missing, unknown, or inapplicable information. While some data models allow representing nulls by special tokens, so-called *disguised missing values (DMVs, in short)* are also frequently encountered: these are values that are not syntactically speaking nulls, but which do, nevertheless, denote the absence, unavailability, or inapplicability of the information. In this work, we tackle the detection of a particular kind of DMV: texts freely entered by human users. This problem is not tackled by DMV detection methods focused on numeric or categoric data; further, it also escapes DMV detection methods based on value frequency, since such free texts are often different from each other, thus most DMVs are unique. We encountered this problem within the ConnectionLens [6–8,12] project where heterogeneous data is integrated into large graphs. We present two DMV detection methods for our specific problem: (*i*) leveraging Information Extraction, already applied in ConnectionLens graphs; and (*ii*) through text embeddings and classification. We detail their performance-precision trade-offs on real-world datasets.

1 Introduction

Digital data is being produced and reused at unprecedented rates. Large datasets are usually processed within data management systems, which *model* the data according to a given data model, provide means to *store* it, and *query* it using a query language. The database industry has been pioneered by Relational Database Management Systems (RDBMSs), whose foundations lay in first-order logic, formalization by E. F. Codd [13], and subsequent work, e.g., [3].

Where There is Data, There are Null Values. Since the early database days, *nulls* have been identified as a central concept denoting missing, unknown, or inapplicable information. The semi-structured data model first embodied in OEM (the Object Exchange Model), proposed to simply omit missing values

© Springer-Verlag GmbH Germany, part of Springer Nature 2022
A. Hameurlain et al. (Eds.): *Transactions on Large-Scale Data- and Knowledge-Centered Systems LI*, LNCS 13410, pp. 97–118, 2022.
https://doi.org/10.1007/978-3-662-66111-6_4

from the data [17]. However, standard semi-structured models such as XML and JSON re-introduced nulls, e.g., xsi:nil in tools supporting XML Schema or the special null value in JSON. Such null tokens may have been needed in practice because perfect, complete databases, regardless of their data model, are the exception rather than the norm.

Disguised Missing Values. In practice, relational databases often feature not only *null* tokens but also *non-null values playing the semantic role of nulls*, also called *disguised missing values (DMVs, in short)* [2]. For instance, 0 or -1 are often used to encode an unknown (but non-zero) number such as a price; users may enter "none", "-", "unknown" or "N/A" or any other similar phrase or token, in entry forms requiring numbers, names or dates that they are unable or unwilling to fill in. Further, when a value needs to be chosen from a predefined set, such as a state of the U.S., users may forget to set it from the menu, leaving the default value which just happens to be the first, e.g., "Alabama" as a U.S. state.

Data entry forms sometimes prevent DMVs by checking the entered value, e.g., "N/A" would not be accepted as a number. However, DMVs may still persist: (*i*) users replace "N/A" with 0 for an unknown, non-zero number; (*ii*) if the expected input type is free text, e.g., "List of industrial collaborations in connection with this research", no simple format-driven validation applies; (*iii*) in cases such as "Alabama" above, the value is in the correct domain.

Detecting DMVs. Identifying DMVs requires dedicated methods, and several have been proposed for relational databases [19,23–25], based on a *statistical analysis* of the data. They can detect, for instance, when a value such as 0 is suspiciously frequent in a numeric attribute, or when a value of attribute $R.a$ is an outlier in the joint distribution of $(R.a, R.b)$, where we expect the distributions $R.a$, $R.b$ to be independent. More approaches, in particular *rule-based*, are discussed in [1]. Other works also propose corrections for erroneous or missing values, e.g., [22]. Detecting and correcting DMVs is important for data cleaning (users may want to replace them with explicit nulls), and for query correctness: null values should not match any selection predicate, and there should be no join on null values.

Problem Statement and Outline. In this work, we consider the detection of DMVs in *textual, heterogeneous data*. The motivation for our work came from ConnectionLens [7,8,12], a system that integrates structured, semi-structured, or unstructured data into graphs, enriched by adding all the entities (people, organizations, places, URIs, dates, etc.) encountered in various text nodes.

We have developed ConnectionLens inspired by fact-checking and data journalism applications [6,9]. In such applications, we encountered many datasets where some fields are *free-form text* entered by users. For instance, in the French Transparency dataset HATVP[1], elected officials need to state "their direct financial participation in company capitals"; in the PubMed bibliographic database[2],

---

[1] https://hatvp.fr.

[2] https://pubmed.ncbi.nlm.nih.gov/.

free-form texts include the article titles, abstracts, funding statements, and possible acknowledgments. Detecting DMVs in ConnectionLens graphs is interesting from the following perspective: if we know that a string is a DMV, we can avoid extracting entities from it, thus reducing the time needed to construct ConnectionLens graphs [6,7].

DMVs encountered in free-form text fields can be classified into two broad classes:

1. Short, simple strings such as "N/A" or "-", which tend to be frequent, and thus can be detected using previously proposed methods, such as [23,25];
2. Complex phrases such as "Liz Smith has not received any funding related to this work", "No conflicts of interest, financial, or otherwise are declared by the authors", or "There is no conflict of interest relating to Authors. The manuscript was prepared according to scientific and ethical rules".

To detect DMVs of the second form above, our work is organized as follows:

1. We recall the closest existing DMV detection techniques (Sect. 2), then we introduce a motivating example to support our discussion (Sect. 3).
2. We show that ConnectionLens Named Entity Recognition can be leveraged to (manually) establish an *entity profile* for each set of text attributes in which we want to detect DMVs and consider any value deviating from this profile a DMV. For instance, the entity profile of financial participation descriptions could be *Organization+* to state that it should contain at least one organization. Profiles allow defining, in semistructured and heterogeneous graphs, groups of values on which we can reason about DMVs. This method is quite accurate, however, it incurs a high computational cost, since entity extraction is a complex operation (Sect. 4).
3. To address this shortcoming, we devised a novel method, which relies on *text embeddings and classification*, while also leveraging entity extraction on a much smaller portion of the dataset. This method is much more efficient than the one based on entity profiles, while also being very accurate (Sect. 5).
4. We show how we *integrated this novel method within the architecture of ConnectionLens* (Sect. 6) to speed up graph construction.
5. In our experimental evaluation (Sect. 7): (*i*) We perform a set of experiments on state of the art methods, and show that they do not perform well on the DMVs we target in this work. (*ii*) We study the efficiency and precision of our method based on embeddings and classification. (*iii*) We experimentally validate the interest of including it within the ConnectionLens system, demonstrating that it reduces the graph construction time.

This work is an invited extension of a short, informally presented paper [10]. A core contribution of this paper with respect to that prior work is the integration of our DMV detection method within ConnectionLens (Sect. 6), together with the associated experiments (Sect. 7.7) validating the practical interest of this method. Other novel extensions, beyond improving and clarifying the writing, include comparisons with a method based on sentence-Bert [27] (Sect. 7.5), and a validation of the quality and performance of our methods on a manually labeled dataset (Sect. 7.6).

2 Related Work

In this section, we recall the main DMV detection methods, as well as some closely related efforts.

Foundational work in this area was made in [24], which introduced and formalized the problem of Disguised Missing Values (DMV), and measured its influence on different data science models. A set of *statistical models* (mutually disjoint hypothesis) were introduced in [21] to model nulls as well as disguised missing values:

- The **MCAR** (Missing Completely At Random) model posits that the probability of a value to be missing is the same for any value of an attribute, and does not depend on the values of any other attribute. For instance, assuming an attribute is the result of a physical measure made with a device that breaks down, the resulting missing values are not correlated to any other aspect of the data or of the values.
- The **MAR** (Missing At Random) model considers that the probability for a value being missing depends on values encountered in other attributes of the same table (these notions have been defined for tables). For example, in a political poll, assume *young* voters are more likely not to declare their political preference. Then, the political preference value is MAR.
- **MNAR** (Missing Not At Random) applies when neither MAR nor MCAR holds, and when the probability for a value to be missing does depend on the actual value that is missing, but not on the values of any other attribute. For instance, assuming supporters of a certain political party generally avoid stating their preference and instead let that information go missing, such values are MNAR.

Building upon these models, [19] has proposed a heuristic method for identifying DMVs in relational databases. Under the MAR and MCAR assumptions, the authors assume that a value v in attribute A_i in a table T is a DMV if $\sigma_{A_i=v}(T)$ contains a subset $T^*_{A_i=v}$ that represents a good sampling of T. Such a subset is an *Embedded Unbiased Sample* (**EUS**) which means that except for attribute A_i, $T^*_{A_i=v}$ and T have similar distributions. Then, a **MEUS** (*Maximal EUS*) is intuitively an EUS with a good trade-off between size (larger is better) and similarity (in distribution) with T. Thus, the MEUS is the largest EUS with the highest similarity. The gist of the [19] heuristics is to find MEUS in a dataset and consider their associated $A_i = v$ values as DMVs.

FAHES [25] incorporates the method of [19], to which the authors add two other methods, in order to distinguish three classes of DMV.

- The first class contains **syntactic outliers**. A syntactic outlier is a value whose syntax is significantly different from that of other values in the same attribute. Two techniques are used to identify them. (*i*) Syntactic pattern discovery infers a frequent syntactic pattern (shape) for the values of each attribute and points out the values that do not fit the pattern as syntactic outliers. For example, if the attribute is "blood type", the recognized pattern could be *one or two uppercase letters followed by a + or a - sign*; then, "ABO" would be considered a syntactic outlier. (*ii*) Repeated Pattern Identification singles out values that contain repeated patterns, such as 0101010101 in a 10-digit phone number, or "blablabla" in a text attribute.
- Second, in numerical attributes, **statistical outliers** can be found by leveraging common outlier detection methods [18]. This allows identifying as DMVs numerical values that do not fit the extent of the other values, e.g., negative values in a distance attribute.
- Finally, some **inlier DMVs**, called **Random DMVs** in [25] can be identified. These are legal attribute values, which do not stand out as outliers; they are the hardest to find even for an application domain specialist. The "Alabama" example from Sect. 1 is a typical example. Inlier DMVs are detected in [25] under the MAR and MCAR hypotheses; the authors state that detecting DMVs under the NMAR model is hard to impossible. The intuition being exploited is that *DMVs are frequent values* (because the lack of information is assumed to occur more frequently than an actual, correct value). Thus, one must find amongst the most frequent values, which ones are DMVs. To this purpose, each frequent value is successively replaced by an actual null. If, by doing this, the (original and introduced) null values follow the MAR or MCAR models, then we consider that value as a good DMV candidate. Then, the MEUS method from [19] is applied to each candidate to detect the DMVs.

The FAHES team has developed a tool using these methods to detect all three types of DMV in a relational database.

DMV detection is also related to *data error detection and correction*, which has been studied in [1]. A tool called RAHA [23] has been developed for detecting errors in relational databases; it detects the DMVs that FAHES [25] finds, as well as errors that are not DMVs. BARAN [22] corrects data errors through a combination of multiple techniques. The free-text DMVs we are interested in are not errors, and their values should be preserved as such. From this perspective, FAHES [25] is closest to our DMV detection goal.

DMV detection is one among the many problems raised by poor *data quality* problems that have been traditionally addressed through *data cleaning*. Data quality raises many real problems, which to this day still need solutions. Traditional approaches for data cleaning were rule-based [15,16,26]. Newer techniques are now based on machine learning, e.g., [20]. DMV detection can also be considered as part of data profiling [2] since DMV detection also allows the characterization of a certain attribute (set of nodes) by the percentage of their values which are DMVs.

Summing up, the existing literature has proposed methods for detecting DMVs and errors, in categorical or text data, with rule-based and especially statistical and more recently machine learning techniques. Our focus is on DMV occurring in free-text data, which are not detected by methods based on value frequency, as we illustrate below.

3   Motivating Example

As part of a data journalism project [6,8], we loaded 400.000 **PubMed bibliographic notices** in a ConnectionLens graph, out of which we extracted (paper ID, **conflict of interest** statement) pairs. These conflict of interest (**CoI**, in short) statements cover any kind of benefits (funding, personal fees, etc.) that authors report with various organizations such as companies, foundations, etc.

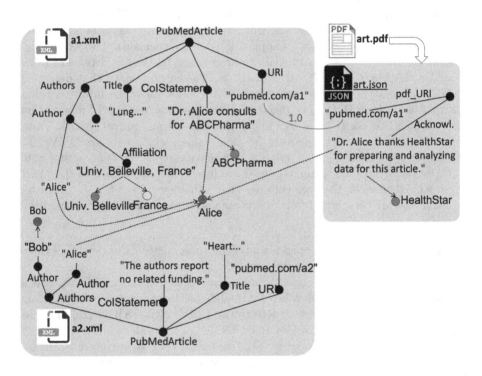

Fig. 1. Sample ConnectionLens graph.

In Fig. 1, "Dr. Alice consults for ABCPharma" (in the upper left) is such a conflict of interest, part of the XML bibliographic notice; "Dr. Alice thanks HealthStar... this article" (at the top right) is another one. "The authors report no related funding" in the second paper, at the bottom of the figure, illustrates the DMVs targeted in this work. PubMed data originates from various biomedical journals. Some do not provide CoI information; in this case, the CoI is an empty string. Others provide a default DMV, e.g., "The authors report no conflict". Finally, some journals only allow free text, leading to a large variety of disguised nulls.

ConnectionLens extracts **named entities** from all text nodes, regardless of the data source they come from, using trained language models. In the figure, blue, green, and yellow nodes denote Organization, Person, and Location entities, respectively. Each entity node is connected to the text node it has been extracted from, by an extraction edge, which also records the confidence (between 0 and 1) of the extractor. Finally, nodes are compared to find that some may be equivalent, or similar.

Entity extraction is a costly operation since it involves predicting, for each token encountered in a given text, if it is part of an Organization entity, Person entity, Location entity, or none. The prediction is made by computations that involve a large, trained model. Trying to extract entities from a DMV is a computational effort spent for no benefit.

4 Detecting DMVs with Entity Profiles

Inspecting the ConnectionLens graph illustrated in Fig. 1, together with our journalist partner, we immediately made the following observation. An actual CoI (such as those involving Alice in Fig. 1) is either of the form "Researcher A was funded by B", or of the form "The authors acknowledge funding from C". Thus, a person's name may be present (in other cases, we just find "The authors"), but an organization is always involved. Thus, we can say that the *entity profile* of a CoI is: it must contain at least an organization.

This leads to the following DMV detection method:

- Extract all named entities from the CoI strings (through regular Connection-Lens data ingestion);
- Declare those CoI strings in which no Organization entity was found, as DMVs.

One could also call such DMVs *uninformative answers*. **We use the result of the above entity-based DMV detection method as a ground truth in our work**, for several reasons: (*i*) Named Entity Recognition is by now a relatively well-mastered task, thus its precision is quite good, and considered acceptable by our end-users; (*ii*) Constructing an ideal human-authored ground truth would require time or monetary costs out of reach for our setting; (*iii*) The very purpose of our work is to save Named Entity extraction time, in other words: Named Entity extraction at the core of ConnectionLens' integration is a given in our context.

The accuracy of this method is exactly that of the entity extractor; it has been shown in [4,5] (for English) and in [7] (for French) that the accuracy is quite high. Its drawback is that *extracting entities from all CoI strings is very lengthy*.

This motivates the search for a faster technique, which, on one hand, could identify the DMVs, while at the same time also reducing the entity extraction (thus, the actual ConnectionLens graph creation) time.

5   DMV Detection Through Embedding and Classification

Our initial DMV detection approach, which does not require extracting named entities from all strings, has been to *cluster* the text values from our motivating example dataset, in order to obtain DMV cluster(s) separated from non-DMV clusters. In particular, we experimented with the K-means [18] algorithm, setting the number of clusters to 10 which was the optimal number of clusters determined using the elbow method[3]. Some clusters contained a majority of DMVs and others a majority of CoIs. However, the clustering was not very successful at separating DMVs from meaningful CoIs.

Thus, we looked for an alternative method. Our idea is: we could extract entities from a small part of the data, so we have a small *automatically* labeled dataset to train a Machine Learning model to recognize DMVs (based on the method described in Sect. 4). We could then use this model to predict whether a yet-unseen value is a DMV or not. Such a model could detect DMVs faster than using the entity extraction technique.

5.1   Textual Data Representation

Many techniques can be used to represent textual data. Transformers like BERT [14] have been proven really efficient for many Natural Language Processing (NLP) tasks; Sentence-BERT [27] provides sentence-level embeddings.

An alternative, less costly textual data representation can be computed by first applying a set of common text pre-processing steps: suppressing punctuation, normalization, and stemming. Embeddings can then be computed using the well-known TF-IDF (Term-Frequency - Inverse-Document-Frequency) representation, also frequently used in NLP. TF-IDF weights term frequencies in each document according to the frequency of the term across the corpus. If a word occurs many times in a document, its relevance is boosted (TF part of the score), as this word is likely to be more meaningful. Conversely, IDF stands for the fact that if the word appears frequently in many documents, then it is probably frequent in any text, and its relevance should be decreased. Finally, to reduce the dimensionality of the representation, we only consider the terms having the 20.000 highest TF-IDF scores.

We will compare Sentence-BERT representation with TF-IDF representation for this task of detecting DMVs in free-form texts. However, we integrate into ConnectionLens (Sect. 6.2) the TF-IDF version, because it is faster yet provides equivalent quality.

---

[3] https://en.wikipedia.org/wiki/Elbow_method_(clustering).

5.2   Classification Model

To classify texts as DMVs or non-DMVs, we decided to rely on a Random Forests classifier [11]. These classifiers are not the fastest, but they are quite efficient over complex data. Random Forests rely on decision trees, which seemed appropriate in our context, as they could learn specific discriminating words that help to differentiate the classes. Other classifiers might work as well; our goal here is to investigate whether the above approach, which trains the classifier on extraction results, can provide a more efficient DMV detection method, by avoiding extracting entities from a certain part of the input.

As we show in Sect. 7.5, this is indeed the case; even for small training set sizes (that is, even if entities are fully extracted only from a small part of the data), the classifier learns to predict DMVs quite accurately, while sparing significant entity extraction time comparing to the entity profile method.

5.3   Placing Our DMV Detection Methods in Context

To further clarify the relationship between our work and prior DMV detection work, Fig. 2 depicts known types of DMVs as rectangles carrying white titles, and the values which various techniques find to be DMVs, as ovals with black titles. As no method is perfect, rectangles and ovals only partially overlap; further, some DMVs are detected by more than one method, and thus some ovals overlap. We assigned numbers to some areas in the figure, and comment on them below. It helps to keep in mind that an area in a rectangle but outside of an oval comprises DMVs that the method corresponding to the oval does not detect; conversely, an area within an oval but outside of the DMV rectangle(s) is a declared by the method to be a DMV, while it is not.

1. Correctly detected statistical outliers, e.g.: In a human height attribute, a height of 3 m.
2. Wrongly detected statistical outliers, e.g.: In a dataset containing salaries of the employees of a company, if the CEO's salary is 10 times higher than any of his employees, this could be wrongly detected as a DMV.
3. Correctly detected syntactic outliers, e.g.: In a blood type attribute, the value 'ABO'.
4. Wrongly detected syntactic outliers, e.g.: In a name attribute, *François-Noël* which is a composed name is detected as a syntactic outlier because of the '-', even though it is a valid name.
5. Correctly detected Random DMVs, e.g.: A default value such as *Alabama* for a state, detected thanks to the MEUS technique.
6. Correctly detected Random DMVs found by the MEUS technique, and also by syntactic and statistic outliers detections. A DMV can be at the same time detected as a syntactic outlier, a random DMV, and a statistical outlier, e.g., a default distance value of -1.
7. Wrongly detected random DMVs. In a poll where we ask for favorite colors, *blue* might come often. Lacking correlation with other attributes, *blue* could be wrongly detected as a random DMV.

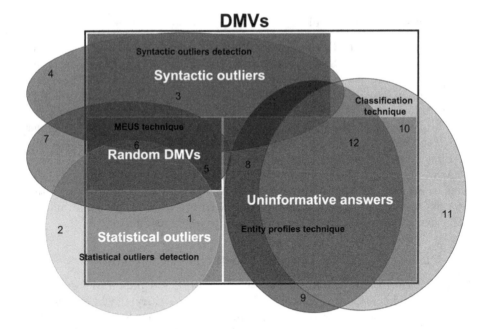

Fig. 2. DMVs and the scope of each DMV detection technique.

8. Correctly detected DMV with entity profile technique, e.g.: In a conflict of interest paragraph: *"The authors report no conflict of interest."*
9. Wrongly detected DMV with entity profile technique. These are errors of the entity extractor when it misses an organization.
10. Correctly detected DMV with the classification method that was not detected by the entity extraction method. For example, in *'John McDonalds declares no conflict of interest'*, the entity extractor could detect McDonalds as an Organisation, and classify the value as an actual CoI instead of a DMV.
11. Wrongly detected DMV with the classification method, these are errors of the classifier.
12. Correctly detected DMV with both classification and entity profile technique. Most of the DMVs detected by the entity profile technique are as well detected by the classification technique.

With respect to the diagram in Fig. 2, we *target* the Uninformative answers rectangle, with *two methods*: the Entity profile one, corresponding to the blue oval that encompasses the areas numbered 8, 9, and 12; and the Classification one, corresponding to the green oval, that comprises the areas numbered 10, 11 and 12. The closest method from the literature, FAHES, excels in finding: Syntactic outliers, Random DMVs, and Statistical outliers.

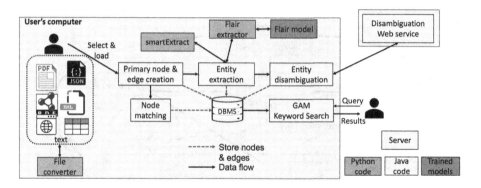

Fig. 3. ConnectionLens architecture (based on [7]).

6 Integrating our DMV Detection Methods within ConnectionLens

We recall the architecture of the ConnectionLens system (Sect. 6.1), then discuss how we integrated our DMV detection module within it (Sect. 6.2).

6.1 ConnectionLens Architecture

The architecture of the ConnectionLens software platform is outlined in Fig. 3. To consolidate heterogeneous data sources into a single graph, the sources are traversed, *creating primary nodes and edges* which represent the source contents (black solid nodes and edges in Fig. 1). On the fly, also during the data source traversal, all the text values encountered in the data are sent to an *entity extraction* module, which may recognize named entities within these texts. Extracted entities are shown as colored nodes in Fig. 1, and edges connecting them to their parent nodes are called *secondary edges*. ConnectionLens' entity extraction (for French and English) is based on Flair [4]. The cost of an extraction operation is relatively high compared with other kinds of processing taking place in memory on a data item, and quite high also when compared to the cost of storing data persistently on disk [7].

Within the ConnectionLens platform, the Flair entity extractor is deployed as a **Python Web service** using the **Flask** web service library; we were able to experimentally check that the performance overhead of calls from the main Java software to the Python web service was negligible. Depending on how many text nodes a data source contains, and how long they are, entity extraction may take a very large part of the graph construction time. A method we implemented to alleviate to some extent speeds up entity extraction by exploiting the parallel processing (multi-core) capabilities of the server on which the extraction runs. Concretely, a **batch** mechanism is implemented: text nodes accumulate in a fixed-size batch and the Flask service is called when the batch is full. It is more efficient for a multi-core machine to apply its prediction model to a set of values

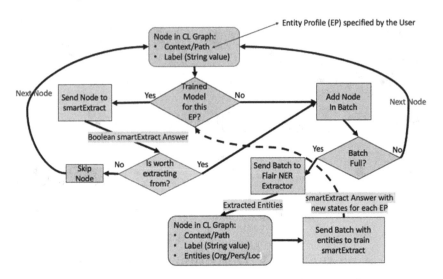

Fig. 4. Integration of DMV detection module within ConnectionLens.

in parallel, as enabled by sending them in batches; the benefits range from 2× on a 4-cores laptop to 20× on a powerful server [7]. However, entity extraction costs remain quite significant, and in some cases, they dominate the ConnectionLens graph construction time.

The remaining ConnectionLens modules are: a *File converter*, that allows ingesting document formats such as PDF, Word, Excel, etc. by converting them to JSON HTML, and/or RDF, which can be directly loaded; a *Node matching* module that creates equivalence or similarity links between entities (red, respectively, dashed graph edges in Fig. 1); an *Entity disambiguation* service, which, for each entity identified in the graph, attempts to find its URI in a knowledge base; finally, a *Keyword search algorithm (GAMSearch)* [6,7] which enables users to search for information within ConnectionLens graphs.

6.2   Integrating Our DMV Detection Method Within ConnectionLens

We now describe how we integrated our DMV detection approach described in Sect. 5 within ConnectionLens, to speed up graph construction. In a nutshell: we train our classifier during the graph construction process, then, in the same process, we use the classifier to predict which strings are DMVs, and thus, not worth the Named Entity Recognition effort.

Value Contexts. The DMV detection method applies to a set of values describing the same kind of data. Thus, while loading complex-structure data, we need to dynamically form such groups of values. For this, we view each value as occurring in a **context**, based on the structure of the dataset it comes from. On hierarchical data sources (JSON, XML, HTML, etc.), a natural context to attach to a value is the root-to-leaf label path on which the value occurs in a given document. In PubMed bibliographic data, for instance, CoI values appear as text children of the XML elements on the path PubMedArticles.PubMedArticle.CoIStatement. Thus, we pair each value on which extraction could be performed (or which could be a DMV), with its context; all the values in a certain context form a set on which we can learn.

Entity Profiles (EP). The next ingredient we need is *entity profiles* (Sect. 4), spelling which entities we expect to find in values from a certain context that are not DMVs; such a judgment is easily made by a human expert. Thus, currently, for each context where an entity profile is known, we allow users to specify them in a configuration file given to ConnectionLens.

SmartExtract Web Service. We deployed our DMV detection approach as a standalone Python Web service, called **smartExtract** in Fig. 3, and also deployed using Flask. It is called by the Entity extraction module with each (value, context) pair, and works as follows (see Fig. 4). To each context, we associate a state that represents the existence (or not) of a TF-IDF vectorizer and a trained Random Forest model (Sect. 5.2) able to predict, based on the entity profile (EP) associated with that context, whether a value occurring in the context is a DMV (Sect. 5). We consider the model has been sufficiently trained when it has reached a determined quality, measured by its f1-score. The quality to reach is a parameter of the service. For each value (text node) that ConnectionLens finds in that context (top box in Fig. 4), we test this state (diamond box below the top box in the figure):

- *Yes*, the model associated with this EP has been fully trained on values previously encountered in the same context. In this case, we send the value to the smartExtract Web Service, which returns a boolean answer:
    - yes, this value is worth extracting entities from (through the Flair entity extraction service mentioned in Sect. 6.1); in this case, the value is added to the extraction batch;
    - no, this value is not worth the entity extraction effort, since our model predicts that the value is a DMV.
- *No*, the model for this EP has not been fully trained yet. In this case, our module needs to learn more about values in this context (by examining more results of the Flair extraction for values in this context). In this case, the node is added to the extraction Batch.

Once the extraction batch is full, ConnectionLens sends all the batch values to its Flair entity extraction service, to find out the entities contained in each value. We then capture these extraction results and share them with the smartExtract service to train its models (one different model for each entity profile). In turn, smartExtract may answer with the information that the state(s) associated with some context(s) have become true (the respective models are sufficiently trained). ConnectionLens keeps track of the context states on which it bases its decision (test box in Fig. 4).

Performance of Online Learning. In the above integration of smartExtract within ConnectionLens, the smartExtract models are continuously trained with each value sent to extraction (*online learning*). In contrast, in the approach we described in Sect. 5, we trained the model once and for all with the training set, and then just used it to predict which values are DMVs. This difference has several consequences:

1. The training effort of our DMV technique drastically increases in this online integration within ConnectionLens.
2. To keep the TF-IDF representation of the data accurate with respect to all the data seen until a certain point during the graph construction, we need to compute a new vectorization and a new model for each new value.

While this online integration strategy achieves a high quality of data representation and prediction, it is computationally very expensive. Indeed, the cost of recomputing the vectorization and repeatedly re-training the model increases quickly with the number of nodes.

To reduce this high computational cost, we initially computed a new vectorization only when re-training the model (after each batch of values sent to the extractor). This strategy allowed us to save some time compared to the precedent, however, we were still losing time (when our goal is to save it) by using the smartExtract Web Service.

Next, we decided not to re-train the model after each batch, but to test it on each batch and re-train only if the quality of the model (as measured by the f1 prediction score) is under a determined threshold. This also brought some modest savings. However, using the smartExtract Web Service was still more computationally expensive than running the entity extractor on all values.

The change that actually brought visible performance benefits was to re-engineer the smartExtract service so that it does not load its model each time a prediction is made. Instead, we kept it in memory on the server side, which significantly reduced the time needed to call the smartExtract service and allowed it to speed up the ConnectionLens graph construction.

7   Experimental Evaluation

In this section, we experimentally study several aspects related to the DMV methods that we have considered in the previous sections. After describing our datasets (Sect. 7.1) and experimental settings (Sect. 7.2), we show how FAHES performs on these datasets and that there is room for improvement in Sect. 7.3. Then, Sects. 7.4 and 7.5 study DMV detection through entity profiling and through embedding (TF-IDF embedding and sentence-BERT embedding) and classification, respectively. In Sect. 7.6 we detail the results of our methods and of FAHES, over a small dataset manually labeled. Sect. 7.7 studies DMV detection impact in ConnectionLens extraction time.

7.1   Datasets

To conduct our experiments, we have built three ConnectionLens graphs out of real-world datasets. These were datasets Le Monde journalists suggested we should integrate in ConnectionLens, for applications including that described in [6,8], and which inspired the research described in this paper. We believe these datasets are representative of many Open Data that is published through government and scientific transparency initiatives: data involving names of people, organizations, and partially obtained by asking individual to fill forms including free text fields.

1. We have loaded the most complete **HATVP XML** transparency dataset (35MB), with data about 270.000 people, in a ConnectionLens graph. From this, we extracted the *montant* (monetary amount) fields which appeared to contain many DMVs[4].
2. We loaded a smaller **HATVP CSV** dataset (2.1 MB), containing information about 9.000 people; this dataset is relational-looking, which simplifies processing it through FAHES.
3. We loaded 400.000 **PubMed bibliographic notices** in a graph, out of which we extracted (paper ID, conflict of interest statement) pairs. These CoI statements cover any kind of benefits (funding, personal fees, etc.) that authors report with various organizations such as companies, foundations, etc., as illustrated in Fig. 1. PubMed data originates from various medical journals. Some do not provide CoI information; in this case, the CoI is an empty string. Others provide a default DMV, e.g., "The authors report no conflict". Finally, some journals only allow free text, leading to a large variety of DMVs.

We will denote these datasets as **DS1**, **DS2**, and **DS3**, respectively. In practice, of course, we only extract entities once from each *distinct* string. It turns out that DS3 had a high number of duplicates (especially some very popular disguised nulls). Removing duplicates led to a new dataset we denote **DDS3** consisting of 82.388 values.

7.2 Settings

All experiments were performed on a MacBook Pro 16 in. from 2019, with a 2.4 GHz Intel Core i9 8-core processor and 32 GB 2667 MHz DDR4 memory. Experiments using sentence-BERT (Table 2) had to be run on Google Colab[5] because of the MacBook's lack of support for CUDA. For consistency, all results in Table 2 are obtained in Google Colab, with a Tesla-T4 GPU. We used ConnectionLens[6] to build the graphs, including in particular the extraction of named entities using Flair [4,5], which we had retrained for French [7]. ConnectionLens graphs are stored in Postgres 9.6; experiment code was written in Python 3.6.

Precision, Recall, and F1 Measures. For our purposes, given that we aim to detect a certain form of DMV, a "positive" example is a DMV, while a "negative"

---

[4] The transparency entry forms require filling in the worth of participations or ownerships in various companies; companies that have closed or did not make benefits, or only have a pro-bono activity, lead to DMVs.
[5] https://colab.research.google.com/.
[6] Available from https://gitlab.inria.fr/cedar/connectionlens.

Table 1. Entity extraction method times

Values	Total characters	Extraction times (s)
500	163.203	56
5.000	1.604.141	669
10.000	3.281.345	1.320
15.000	5.000.364	2.175
20.000	6.728.493	2.620

example is an informative value. This interpretation is used in all the experiments described below. In the context of our project, *precision is more important than recall*, since low precision means wrongly considering a value as a DMV, while it contains valid information, which journalists would not want to miss. In comparison, low recall "only" means that we would waste time extracting entities from DMVs that turn out not to contain interesting information. While undesirable, the impact, from an application perspective, is lower.

7.3 DMV Detection Through FAHES

We have applied FAHES [25] on the three datasets described previously, asking it to detect DMVs. It is worth mentioning that FAHES is an unsupervised method; we use it for comparison as its goal of DMV detection is closest to our work. We comment on its results below.

Results on DS1. Among the 270.000 values of the numeric *amount* attribute, FAHES correctly found the 0 (which occurs 45.000 times) as a Random DMV (Inlier). It also detected 372.2196 (4 occurrences) as a numerical outlier DMV; this is wrong. All the amount values are numbers, and as far as we could see, there are no other DMVs.

Results on DS2. In this relational dataset, in an attribute called `filename`, FAHES identified **correctly** the DMV *dispense* (120 occurrences) as a Random DMV. Then, FAHES identified **wrongly** other values as being DMVs, in all cases as Syntactic Outliers DMV. The values falsely flagged as DMVs are:

- *François-Noël* (3 occurrences) in the attribute `given name`;
- *BÉRIT-DÉBAT* (6 occurrences) and *KÉCLARD-MONDÉSIR* (3) in the attribute `name`;
- *di* (4480 occurrences) in the attribute `document type`; this is the acronym for *déclaration d'intérêt*;
- *2A* and *2B* as departement numbers; they are, in fact, correct numbers of French departments in Corsica;
- four distinct, correct URLs within the `photo_url` attribute, probably because their structure did not resemble the others'.

DS2 seems to include no other DMVs.

Results on DS3. Out of the 400.000 values, FAHES correctly identified *"The authors have declared that no competing interests exist"* (31.891 occurrences) as a Random DMV. However, visual inspection exhibited many other DMVs (we will revisit this below). FAHES fails to find them because freely written texts rarely coincide, thus FAHES' statistical approach based on value frequencies leads it to consider a rarely occurring DMV as a valid value, which is wrong. All DMVs detected by FAHES on this dataset are actual DMVs (precision of 1.0). However, in this example, the entity profile technique identifies 351.123 DMVs, most of which FAHES misses, leading to a recall of 0,0909. *This low recall shows the need for alternative techniques for detecting the DMVs we are interested in: uninformative, free-text answers.*

From these experiments, we conclude that FAHES fails to detect the DMVs we are interested in; DMVs detected as Syntactic outliers are often false positives. With a bit of domain knowledge, it is possible to manually discard these DMVs. However, importantly, FAHES has also shown its limitations on uninformative free text DMVs, by missing a vast majority of them. This shows the need of dedicated methods to detect such DMVs.

7.4   DMV Detection Through Entity Profiles

To measure performances of the entity profiles technique, we performed 5 experiments with respectively the 500, 5.000, 10.000, 15.000, and 20.000 first values of dataset the duplicate-free dataset DDS3. The objective here is to measure the extraction time as a function of the input size; this indicates the time needed to extract DMVs using the Entity Profile method. We present the results in Table 1. We observe that extracting entities is time consuming and that the extraction time is almost linear to the number of values. For the complete dataset DDS3, we can expect to have an extraction time of around 11.000 s (183 min), which is quite lengthy.

7.5   DMV Detection Through Embedding and Classification

The most common training-test split method consists on separating the dataset with 20% used for training and 80% for testing. We know that in our case, the most time-consuming operation is to label the training set with the entity extraction technique. Thus, to gain time, we want to reduce the training set.

To evaluate the impact of the training set size on the performance of the model used to detect DMVs over DDS3, we have performed three experiments. We trained models with respectively 20% (16.477 values), 10% (8.238 values), and 1% (823 values) of the dataset and report the comparison of the performances of each model in Table 2. In this table, the precision, recall, and thus F1 are computed using the result of the entity profile method (Sect. 4) as the gold standard. In bold, we highlight the best performance result between TF-IDF and sentence-BERT. Table 2 shows that we can attain very good precision, even if the model is trained on a small part of the dataset while saving significant amounts of time. Table 2 also shows that using sentence-BERT to represent our

Table 2. Impact of the training set size on the performance of DMV detection.

Training-set size	16.477		8.238		823	
Embedding type	TF-IDF	s-BERT	TF-IDF	s-BERT	TF-IDF	s-BERT
Embedding Times (s)	99	509	99	509	99	509
Extraction Times (s)	2.153	2.153	1.075	1.075	108	108
Training Times (s)	20	75	10	30	2	1
Prediction Times (s)	4	2	4	2	3	1
Total Times (s)	**2.276**	2.739	**1.206**	1.616	**212**	619
Precision	0,939	**0,956**	0,933	**0,956**	0,926	**0,946**
Recall	**0,922**	0,877	**0,923**	0,870	**0,872**	0,856
F1-score	**0,931**	0,915	**0,928**	0,911	**0,899**	**0,899**

Table 3. Comparing FAHES, Entity profile and classification over a sample of 200 manually-labeled values.

Techniques	TP	FP	TN	FN	Precision	Recall	F1-score
FAHES	0	0	96	104	0	0	0
Entity profile	82	4	92	22	0,953	0,788	0,862
Classification TF-IDF	94	2	94	10	0,979	0,904	0,940

data does not significantly improve performance, while it heavily increases execution time. This happens despite speeding up its execution as much as possible, by parallelizing execution over batches of 1.000. Interestingly, Table 2 also shows that in our problem, precision is less sensitive than recall to the reduction of the training-set size which suits our purpose.

With respect to our motivating example, building the graph for DDS3 took around 11.000 s. Using our method with a training-set of 1% of the values (823 values) takes now the time to predict to which values we have to apply the extractor (125 s), to which we add the time to extract the valuable values. We have found on our dataset that there are around 45.000 valuable values. We need 5.900 s to extract those. That brings us to a total of 6.000 s **to build our graph instead of** 11.000 s **previously**, without losing much information.

7.6   Comparison on Manually Labeled Data

To get a better understanding of the performance of our entity profile technique, our classification technique, and FAHES, we labeled by hand a sample of 200 values from DDS3. The sample contains 96 actual CoIs, and 104 DMVs, labeled following the instructions of our domain expert (journalist).

Table 3 shows, for each method, the numbers of true and false negatives (TN and FN), as well as the precision, recall, and F1 score of each technique *with respect to the human-labeled gold standard.* Our first observation is that FAHES has not found a single DMV in this dataset, leading to recall, precision, and F1 of 0.

The Entity Profile technique performed quite well (F1 of 0, 86). A few valid CoI statements are wrongly detected as DMVs, but the contrary is quite rare, and that suits our application needs since we should not skip extraction on valid CoIs.

Our classification method performed even better (F1 of 0, 94). Since this method is based on word frequency, we can suppose that some words are specific to DMVs and others are specific to valid CoIs.

Inspecting the false negatives (DMVs detected as valid CoIs), we noticed that they mention "ICMJE", as in *Conflicts of Interest: All authors have completed the ICMJE uniform disclosure form (available at* http://dx.doi.org/10.21037/ atm-20-3650). *The authors have no conflicts of interest to declare.* ICMJE has been wrongly detected as an entity, which led astray our entity profile technique. As our classification model is trained with entity profile results, "ICMJE" has probably been identified as associated with valid CoIs.

On the other hand, the entity profile technique has led to only having 4 CoIs identified as DMVs (false positives). In those, the authors mentioned patents they have, which may be seen as creating a conflict of interest with the research presented in the paper. However, by the domain experts' (journalists') rule, these statements are no conflicts of interest.

Finally, we found that the NER has missed an organization entity in the value *"The authors are supported by the use of resources and facilities at the Michael E. Debakey VA Medical Center, Houston, Texas"*, thus the entity profile detection technique has wrongly considered this string to be a DMV. However, the classification model did not make the same mistake and correctly considered this value an informative COI.

7.7   DMV Detection Integrated in ConnectionLens

As explained in Sect. 6.2, we developed a Python Flask service for DMV detection based in entity profiles, to save extraction time in specified contexts. We only kept TF-IDF representation, because sentence-BERT has proven to be way slower. To measure its impact on the time spent extracting entities, we loaded 100.000 XML PubMed bibliographic notices into ConnectionLens, without the smartExtract service (that is, following the graph construction method prior to this work), then also using the smartExtractor service as described in Sect. 6.2. The quality parameter (F1-score mentioned in Sect. 6.2) is set to 0.9. To mesure the F1-score, we split each batch of values used for training, into training testing subsets, the latter containing 20% randomly chosen values from the batch. We used a batch size of 160, that is: the Flair NER entity extractor was called on groups of 160 values. In this experiment, the smartExtract service is called to ConflictOfInterest values, for which an entity profile is manually provided. Thus, in this experiment, we only apply entity extraction on the ConflictOfInterest values.

As we can see in Table 4, we saved 557 calls to the Flair NER Service (which amounts to 70% of the calls) thanks to the smartExtract service. This corresponds to saving 1.770 s on extraction (33% of the extraction time) and 2.215 s

Table 4. Impact of the smartExtract web service on entity extraction within ConnectionLens.

Is the smartExtract web service used?	Yes	No
Number of calls to Flair NER service	228	785
Time needed for extraction (seconds)	3.514	5.284
Total graph construction time (seconds)	6.865	9.080
Number of organization entities extracted	35.110	38.335

from the total graph construction time. Finally, using our smartExtract service leads to missing out on 3.225 organization entities that would have been extracted without it. These represent about 10% of the total organization entities, which is consistent with our quality parameter (F1-score = 0.9).

The overall time saving is slightly higher than the one corresponding only to extraction. This is because not creating some entities also means we do not have to store them, nor the edges that connect them to their parents in the graph.

8   Conclusion

Integration of very heterogeneous data, such as we encountered in data journalism applications [8,9], requires extracting Named Entities from all values found in the data, in order to interconnect sources through the presence of the same entity in two or more sources. The ConnectionLens [6,7] system we have developed leverages this idea.

In this work, we sought to automatically find *uninformative answers*: these are free texts that users write in response to some question, and which do not provide the information required by the question. These can be seen as a particular class of textual DMVs. While many kinds of DMVs have been studied in the literature, as we discussed in Sect. 2 and 5.3, when uninformative answers are basically all distinct, existing methods cannot detect them.

The first technique we studied leverages ConnectionLens' effort carried to extract Named Entities from each value of the database. Our technique exploits so-called *entity profiles* (expected entities in a valid value) to identify DMVs. While highly accurate, this is expensive time-wise, because of the extraction. For efficiency, instead, our second method trained a Random Forest classifier on a small subset of our dataset, labeled with entity profiles, and classified the other values as DMVs or not. This technique saves significant extraction time, while also having very good accuracy.

We have included this method within the ConnectionLens platform and demonstrated that it allows to avoid a large part of the Named Entity Recognition (NER) effort. This is significant, since as shown in [7], NER dominates the graph construction time. Thus, the method proposed here allows speeding up the construction of integrated graphs out of heterogeneous data sources.

Acknowledgments. This work was supported by the ANR AI Chair project Sources-Say Grant no ANR-20-CHIA-0015-01. Galhardas' work was supported by national funds through FCT under the project UIDB/50021/2020.

References

1. Abedjan, Z., et al.: Detecting data errors: where are we and what needs to be done? Proc. VLDB Endow. **9**(12), 993–1004 (2016)
2. Abedjan, Z., Golab, L., Naumann, F., Papenbrock., T.: Data Profiling. Morgan and Claypool (2020)
3. Abiteboul, S., Hull, R., Vianu, V.: Foundations of Databases. Addison-Wesley, Boston (1995)
4. Akbik, A., Bergmann, T., Blythe, D., Rasul, K., Schweter, S., Vollgraf, R.: Flair: an easy-to-use framework for state-of-the-art NLP. In: ACL (2019)
5. Akbik, A., Blythe, D., Vollgraf, R.: Contextual string embeddings for sequence labeling. In: ACL (2018)
6. Anadiotis, A.-C., et al.: Empowering Investigative Journalism with Graph-based Heterogeneous Data Management. Bull. Tech. Committee Data Eng. (2021)
7. Anadiotis, A.C., et al.: Graph integration of structured, semistructured and unstructured data for data journalism. Inf. Syst. **104**, 101846 (2021)
8. Anadiotis, A.-C.G., et al.: Discovering conflicts of interest across heterogeneous data sources with connectionlens. In: CIKM 2021: The 30th ACM International Conference on Information and Knowledge Management, Virtual Event, Queensland, Australia, November 1–5, 2021, pp. 4670–4674. ACM (2021)
9. Bonaque, R., et al.: Mixed-instance querying: a lightweight integration architecture for data journalism. In: VLDB (2016)
10. Bouganim, T., Galhardas, H., Manolescu, I.: Efficiently identifying disguised nulls in heterogeneous text data. In: BDA (Conférence sur la Gestion de Données - Principles, Technologies et Applications), Paris, France, October 2021. Informal publication only (2021)
11. Breiman, L.: Random forests. Mach. Learn. **45**(1), 5–32 (2001)
12. Chanial, C., et al.: ConnectionLens: finding connections across heterogeneous data sources (demonstration). PVLDB (also at BDA) **11**(12), 4 (2018)
13. Codd, E.F.: A relational model of data for large shared data banks. Commun. ACM **13**, 377–387 (1970)
14. Devlin, J., Chang, M.-W., Lee, K., Toutanova, K.: Bert: pre-training of deep bidirectional transformers for language understanding. In: ACL (2019)
15. Galhardas, H., Florescu, D., Shasha, D.E., Simon, E.: Declaratively cleaning your data with AJAX. In: Doucet, A. (ed.) BDA (2000)
16. Galhardas, H., Florescu, D., Shasha, D.E., Simon, E., Saita, C.-A.: Declarative data cleaning: language, model, and algorithms. In: VLDB (2001)
17. Hammer, J., Garcia-Molina, H., Ireland, K., Papakonstantinou, Y., Ullman, J.D., Widom, J.: Information translation, mediation, and mosaic-based browsing in the TSIMMIS system. In: SIGMOD (1995)
18. Han, J., Kamber, M., Pei, J.: Data Mining Concepts and Techniques. Morgan Kaufmann, Waltham (2011)
19. Hua, M., Pei, J.: Cleaning disguised missing data: a heuristic approach. In: SIGKDD (2007)
20. Ilyas, I.F., Soliman, M.A.: Probabilistic ranking techniques in relational databases. In: Synthesis Lectures on Data Management. Morgan & Claypool Publishers (2011)

21. Little, R.J.A., Rubin, D.B.: Statistical Analysis with Missing Data, vol. 793, 1st ed.. Wiley, New York (1987)
22. Mahdavi, M., Abedjan, Z.: Baran: Effective error correction via a unified context representation and transfer learning. Proc. VLDB Endow. **13**(12), 1948–1961 (2020)
23. Mahdavi, M., et al.: A configuration-free error detection system. In: Proceedings of the 2019 International Conference on Management of Data, SIGMOD 2019, New York, NY, USA, 2019, pp. 865–882. Association for Computing Machinery (2019)
24. Pearson, R.K.: The problem of disguised missing data. SIGKDD Explor. Newsl. **8**(1), 83–92 (2006)
25. Qahtan, A.A., Elmagarmid, A., Fernandez, R.C., Ouzzani, M., Tang, N.: Fahes: a robust disguised missing values detector. In: SIGKDD (2018)
26. Raman, V., Hellerstein, J.M.: Potter's wheel: an interactive data cleaning system. In: VLDB (2001)
27. Reimers, N., Gurevych, I.: Sentence-BERT: Sentence embeddings using Siamese Bert-networks. In: EMNLP (2019)

Digital Preservation with Synthetic DNA

Eugenio Marinelli[1]([✉]), Eddy Ghabach[1], Yiqing Yan[1], Thomas Bolbroe[3], Omer Sella[2], Thomas Heinis[2], and Raja Appuswamy[1]

[1] EURECOM, Biot, France
{eugenio.marinelli,eddy.ghabach,yiqing.yan,raja.appuswamy}@eurecom.fr
[2] Imperial College, London, UK
{osella,t.heinis}@imperial.ac.uk
[3] Rigsarkivet, Copenhagen, Denmark
tbo@sa.dk

Abstract. The growing adoption of AI and data analytics in various sectors has resulted in digital preservation emerging as a cross-sectoral problem that affects everyone from data-driven enterprises to memory institutions alike. As all contemporary storage media suffer from fundamental density and durability limitations, researchers have started investigating new media that can offer high-density, long-term preservation of digital data. Synthetic Deoxyribo Nucleic Acid (DNA) is one such medium that has received a lot of attention recently. In this paper, we provide an overview of the ongoing collaboration between the European Union-funded, Future and Emerging Technologies project OligoArchive and the Danish National Archive in preserving culturally important digital data with synthetic DNA. In doing so, we highlight the challenges involved using DNA for long-term preservation, and present a holistic data storage pipeline that brings together several novel techniques (standardized file storage, motif-based DNA encoding, scalable read consensus to name a few) to provide reliable, passive, obsolescence-free digital preservation using synthetic DNA.

Keywords: DNA storage · Long-term archival · Preservation · SIARD-DK

1 Introduction

Today, we live in an increasingly digital society. Digital data pervades all disciplines and has established itself as the bed rock that drives our society, from enabling data-driven decisions based on machine learning, to encoding our collective knowledge compactly in a collection of bits. Thus, preservation of digital data has emerged as an important problem that must be addressed by not just memory institutions today, but also by institutions in several other sectors.

In order to preserve digital data, it is necessary to first store the data safely over a long time frame. Historically, this task has been complicated due to

© Springer-Verlag GmbH Germany, part of Springer Nature 2022
A. Hameurlain et al. (Eds.): *Transactions on Large-Scale Data- and Knowledge-Centered Systems LI*, LNCS 13410, pp. 119–135, 2022.
https://doi.org/10.1007/978-3-662-66111-6_5

several issues associated with digital storage media. All current media technologies suffer from density scaling limitations resulting in storage capacity improving at a much slower rate than the rate of data growth. For instance, Hard Disk Drive (HDD) and magnetic tape capacity is improving only 16–33% annually, which is much lower than the 60% growth rate of data [10]. All current media also suffer from media decay that can cause data loss due to silent data corruption, and have very limited lifetime compared to the requirements of digital preservation. For instance, HDD and tape have a lifetime of 5–20 years. A recent survey by the Storage and Networking Industry Association stated that several enterprises regularly archive data for much longer time frames [19]. Thus, the current solution for preserving data involves constantly migrating data every few years to deal with device failures and technology upgrades. A recent article summarized the financial impact of such media obsolescence on the movie industry [17].

In project OligoArchive [2], we are exploring a radically new storage media that has received a lot of attention recently—Deoxyribo Nucleic Acid (DNA) [6,9,13,16]. DNA is a macro-molecule that is composed of smaller molecules called *nucleotides*(nt). There are four types of nucleotides: Adenine (A), Cytosine (C), Guanine (G), and Thymine (T). DNA used for data storage is typically a single-stranded sequence of these nucleotides, also referred to as an *oligonucleotide* (oligo). DNA possesses several key advantages over current storage media. First, it is an extremely dense three-dimensional storage medium with a capacity of storing 1 Exabyte/mm^3 which is eight orders of magnitude higher than magnetic tape, the densest medium available today [8]. Second, DNA is very durable and can last millennia in a cold, dry, dark environment. A recent project that attempted to resurrect the Woolly Mammoth using DNA extracted from permafrost fossils that are 5000 years old is testament to the durability of DNA even under adverse conditions [18]. Thus, data once stored in DNA can be left untouched without repeated migration to deal with technology upgrades. Third, as long as there is life on earth, we will always have the necessity and ability to sequence and read genomes, be it for assembling the genome of a previously-unknown species, or for sequencing the genome to detect diseases causing variations. As a result, unlike contemporary storage technologies, where the media that stores data and the technology to read data are tightly interlinked, DNA decouples media (biological molecules) from read technology (sequencing), thus reducing media obsolescence issues.

In this work, we provide an overview of the ongoing collaboration between the Danish National Archive and project OligoArchive in demonstrating a holistic solution for long-term preservation of culturally significant data using DNA. We present a motivating use case for long-term digital preservation, outline the challenges involved in using DNA as a digital storage medium, and present the end-to-end pipeline we have put in place to overcome these challenges.

2   Context and Background

2.1   Danish National Archive Use Case

The Danish National Archives is a knowledge center documenting the historical development of the Danish society. The archive collects, preserves and provide access to original data with the purpose of supporting current and future possible needs of the Danish community - public authorities as well as private citizens. A huge part of this work includes the preservation of digitally created and retro-digitized data securely and cost-effectively. Thus, the archive has received and preserved such data since the 1970s.

The archival material used for this work consists of selected hand-drawings made by the Danish king Christian IV (1577–1648). Although his reign was marked by military defeat and economic decline, Christian IV stands out as one of the most prominent, popular and admired characters in the line of Danish kings. The hand-drawings date to the period 1583–1591 where the king was 6–14 years old. The material is a part of a larger archival unit consisting of numerous documents and records[1]. The specific image[2] used for this experiment presents a naval battle between several warships. Besides emphasizing the young king's admiration for warfare and naval tactics, the material further indicates his high level of cultural education as well as his talent for drawing. At Danish National Archive, the material is thus ranked as having "Enestående National Betydning" (meaning unique national significance).

2.2   DNA Storage Challenges

Using DNA as a digital storage medium requires mapping digital data from its binary form into a sequence of nucleotides using an encoding algorithm. Once encoded, the nucleotide sequence is used to *synthesize* DNA using a chemical process that assembles the DNA one nucleotide at a time. Data stored in DNA is read back by *sequencing* the DNA molecules and decoding the information back to the original digital data.

A simple way to convert bits into nucleotides is to adopt a direct mapping that converts 2 bits into a nucleotide, for instance 00 to A, 01 to T, 10 to G, and 11 to C. This way a binary sequence is translated to an arbitrary sequence of nucleotides. However, such a simple approach is not feasible due to several biological limitations imposed by DNA synthesis and sequencing steps. First, DNA synthesis limits the size of an oligo between hundred to few thousands of nucleotides. Therefore, data must be divided into several pieces, with each piece being stored in an oligo. However, unlike current storage devices, oligos do not have logical addressing. Hence, indexing information that can help to identify the order in which the oligos, and hence the corresponding data bits, must be reassembled back during recovery must be stored together with the data bits and integrated in each oligo.

---

[1] https://www.flickr.com/photos/statensarkiver.
[2] https://www.flickr.com/photos/statensarkiver/28273082238.

Second, oligos with repeat sequences (like ACACAC), or long consecutive repeats of the same nucleotide (like AAAAA), and oligos with extreme GC content, where the ratio of Gs and Cs in the oligo is less than 30% or more than 70%, are known to be difficult to synthesize, sequence, and process correctly. Thus, when constructing oligos, constraints need to be enforced to minimize homopolymer repeats and balance GC content. Further, care must be taken to minimize similarity across oligos as having too many oligos with positionally-similar nucleotide sequences can exacerbate sequencing errors and make it difficult to identify the original oligo.

Third, sequencing and synthesis are not error free even for well-formed oligos, as they introduce substitution errors, where a wrong nucleotide is reported, or indel errors, where spurious nucleotides are inserted or deleted. Both sequencing and synthesis also introduce bias. Some oligos are copied multiple times during synthesis, while others are not. Similarly, some oligos are read thousands of times during sequencing while others are not sequenced at all. Thus, it is important to use error correction codes in order to recover data back despite these errors.

In addition to the aforementioned media-level challenges in using DNA as a digital storage medium, there are also other problems associated with digital preservation that DNA does not solve. Any digital file stored on DNA is an encoded stream of bits whose interpretation makes sense only in the context of the application used to render, manipulate, and interact with that file format. While DNA might be able to store data for millennia, the associated applications and file formats might become obsolete. Thus, in addition to preserving data, it is also necessary to preserve the meaning of data by ensuring that data is stored in a preservation-friendly, non-proprietary format. Digital data can also be altered due to a variety of reasons and additional data-integrity techniques should be put in place to ensure that data retrieved from DNA can be trusted to be the same as the original source. The digital preservation community has long pioneered file formats, information systems, and operational methodologies for solving such format obsolescence issues [1]. Thus, a holistic DNA-based preservation solution should build on such techniques to solve both media and format obsolescence issues.

3   Design

In this section, we will describe the end-to-end pipeline we have put in place to overcome the aforementioned challenges.

3.1   Overcoming Format Obsolescence with SIARD-DK

In Denmark, all public institutions and organizations that produce data worthy of persevering are legally bound to submit them to a public archive. As the vast majority of data in the Danish public sector are organized as databases with or without files in various formats, the focus has been on archiving these data in a standardized, system-independent and cost efficient manner. As a result, the archive has implemented a Danish version of the SIARD format (Software Independent Archiving of Relational Databases) [7] named SIARD-DK for storing of such data. SIARD is an open format, designed for archiving relational databases in a vendor-neutral form and is used in the CEF building block "eArchiving".

The first step in preserving data is extracting it and creating an SIARD-DK Archival information Package (AIP). In the creation of this particular AIP, the digitized material was converted to TIFF format. Information relevant to the images such as the preservation format of the files, their title, creator and original size, descriptive information, etc., was extracted and packaged together with relevant documentation in the AIP-format. This was done using proprietary tools developed at the Danish National Archive. The usage of SIARD for storing the files guarantee that the material is preserved in a rich format with relevant metadata stored in a standardized, system and vendor independent way. The resulting AIP is a single ZIP64 file that internally contains the TIFF images, in addition to XML and XSD files that store the schema of the archive and metadata information. This allows for strict validation of the AIP. Further, an MD5 value of each file is stored inside the archive and serves as the fixity to verify data integrity on retrieval.

3.2   DNA Data Storage Pipeline

The end-to-end DNA media storage pipeline is presented in Fig. 1. In the rest of this section, we will provide an overview of both the write path that takes as input the SIARD zip file and stores it in DNA, and the read path that restores back the zip file from DNA.

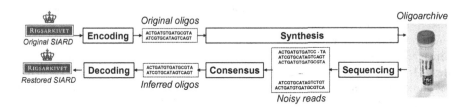

Fig. 1. DNA storage pipeline

Write Path. In order to store the archive on synthetic DNA, the zip file is first encoded from binary into a quaternary sequence of oligonucleotides, and then synthesized to generate synthetic DNA. The steps for encoding the SIARD archive file into oligos is presented in Fig. 2. During encoding, the file is read as a stream of bits and pseudo randomized. In other encoding methods, randomization is used as a way to limit the number of homopolymer repeats in

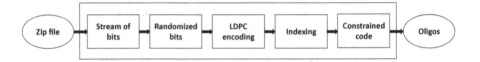

Fig. 2. Encoding bits into oligos

each oligo. In our encoding, homopolymer repeats are handled by an inner constrained code that we explain later. Thus, we do not need randomization for avoiding homopolymer repeats. We use randomization primarily to improve the accuracy of our clustering and consensus methods in the data decoding stage. As mentioned before, data stored in DNA is read back by sequencing the DNA to produce *reads*, which are noisy copies of the original oligos that can contain insertion, deletion, or substitution errors. Our read clustering and consensus methods rely on the fact that the original oligos are well separated in terms of edit distance so that the distance between a noisy read and its corresponding oligo is much smaller than the distance between two oligos. This assumption makes it possible to cluster similar reads and infer the original oligo with a high accuracy. A long sequence of zero or one bits can violate this assumption as they can lead to multiple oligos being similar, or even identical, to each other. A long sequence of bits can also lead to oligos with repetitive sequences (example: ACACACAC). This can pose problems during data decoding, especially if paired-end sequencing is used, where the DNA is partially read from either direction (5" to 3" and 3" to 5") with overlap. In such a case, the two reads corresponding to each orientation must be merged into a single representative read. Repetitive sequences can create issues during this merging process. While we can develop more advanced solutions to perform merging, randomization provides a simple solution that eliminates such issues while ensuring that similarity across oligos is also minimized.

After randomization, error correction encoding is applied to protect the data against errors. We use large-block length Low-Density Parity Check (LDPC) codes [11] with a block size of 256,000 bits as the error correction code, as it has been shown to be able to recover data in the presence of intra-oligo errors, or even if entire oligos are missing [5]. We configure LDPC to add 10% redundancy to convert each sequence of 256,000 bits into 281,600 bits with data and parity. Each 281,600 bit sequence is then used to generate a set of 300-bit sequences, where each 300-bits is composed of 281 data bits and a 19-bit index that is used to order the sequences. Each 300-bit sequence is then passed to a constrained code that converts it into an oligonucleotide sequence.

The constrained code essentially views each oligo as a concatenation of several shorter oligonucleotide sequences, that we henceforth refer to as *motifs*. In our current configuration, the constrained code breaks up each 300-bit sequence into a series of ten 30-bit integers. Each 30-bit integer is fed as input to a *motif generator* that takes the 30-bit value and produces a valid 16nt (nucleotides) motif as output. The motif generator does this mapping by pre-constructing an associative array where the 30-bit value is the key and a 16nt motif is the value. This array is built by first enumerating all possible motifs of length 16nt. Then, all motifs that fail to meet a given set of biological constraints are eliminated.

Our current motif generator is configured to allow up to two homopolymer repeats (AA, CC, GG, or TT), and admits motifs with a G-C content in the range 0.25 to 0.75. With these constraints, using 16nt motifs, out of 4^{16} possible motifs, we end up with only 1,405,798,178 unique, valid motifs with which we can encode any possible 30-bits of data (as the number of all possible 30-bits values is $2^{30} = 1,073,741,824$ which is lower than 1,405,798,178. Thus, we use the billion motifs as values in the array corresponding to keys in the range 0 to 2^{30}. The 30-bit value is thus used as the key into this associative array to produce the corresponding 16nt motif. Thus, at the motif level, the encoding density is 1.875 bits/nt. The reason we limited ourselves to 16nt and 30bits is the fact that this associative array occupies around 100 GB of memory, which we can easily meet using our current hardware. The motif generator can be extended to larger motif sizes and more relaxed biological constraints which can lead to higher bit densities. But as this would require the use of external storage, we leave this open to future work.

Using the associative array, each 300-bit sequence is encoded as concatenation of ten motifs, each with a length of 16nt, leading to an oligo that is 160 nucleotides long. We would like to explicitly point out here that the length of an oligo is a configurable parameter. Thus, while we use 160nts in our current system due to favorable pricing provided by our synthesis provider, our encoder can generate shorter or longer oligos if necessary, and automatically adjust various aspects (like the 19-bit index and 281-bit data size) based on desired oligo length.

Read Path. To retrieve back the SIARD archive, the DNA is sequenced in order to retrieve back the nucleotide sequence of oligos. As mentioned before, sequencing produces reads, which are noisy copies of the original oligos. Thus, we need a consensus procedure to infer the original oligos from the reads. In prior work, we structured this process as sequence of three algorithms, as depicted in Fig. 3. First, we identify all pairs of strings that are similar to each other. As modern sequencers produce hundreds of millions of reads, this first task is extremely computationally intensive due to use of the edit distance as a metric for comparing strings. Thus, we have developed an efficient similarity join algorithm, called OneJoin [15], that exploits the fact that due to randomization during encoding, reads corresponding to the same original oligo are "close" to each other despite some errors and "far" from the reads related to other oligos. The results obtained from the join algorithm are then used to quickly identify clusters of strings that are similar to each other. Each cluster thus groups all reads belonging the same oligo. Finally, we apply a position-wise consensus procedure that uses multiple reads to infer the original oligo in each cluster.

Noisy Reads	Edit Similarity Join	Clustering	Consensus
	1. ACTGTTGTGATCC - TA		
	3. ACTGATGTGATTCAAA	1. ACTGTTGTGATCC - AA	
1. ACTGTTGTGATCC - TA		3. ACTGATGTGATTCAAA	
2. ATCGTACATAATCAGT		5. ACTGAATGTATTAACA	ACTGATGTGATTCAAA
3. ACTGATGTGATTCAAA	2. ATCGTACATAATCAGT		ATCGTACATAATCAGT
4. ATCGTGAATAGCCTGT	4. ATCGTGAATAGCCTGT		
5. ACTGAATGTATGAGTCA		2. ATCGTACATAATCAGT	
	1. ACTGTTGTGATCC - TA	4. ATCGTGAATAGCCTGT	
	5. ACTGAATGTATGAGTCA		

Fig. 3. Various steps in the OneJoin consensus procedure.

While our prior approach was able to infer original oligos with a high accuracy, there were two problems. First, we found that under some datasets, particularly for high coverage reads, OneJoin's memory and computational usage were too high. As OneJoin is a string similarity join, it produces as output all possible pairs of reads that are similar to each other. Thus, given a coverage N, the computational and memory requirements of OneJoin were $O(N^2)$. Second, the use of a general purpose string similarity join led to functionality repetition at multiple places in the read path. For instance, OneJoin internally uses an edit distance check to filter out strings that are not similar. Later in the read pipeline, we had to repeat the edit distance computation in the alignment stage once to get position-wise consensus. This repetition led to needless overhead. To solve these problems, we have developed a new consensus procedure that we refer to as *OneConsensus*.

In the following sections, we describe the key stages of OneConsensus algorithm. In order to efficiently identify and group all the similar reads, our algorithm relies on two well-known algorithmic tools that allow to drastically cut down the computational time: CGK-Embedding and Locality Sensitive Hashing (LSH).

CGK Embedding. As we mentioned in the previous section, the similarity metric used in OneConsensus is the edit distance. Given two strings x and y, the edit distance is defined as the minimum number of edit operations i.e. insertions, deletions and substitutions, necessary to transform x in y. Another metric, commonly used to compare strings is the Hamming distance. However, the latter takes into account only the number of mismatches between the two strings, or in other words the number of substitutions to transform x in y. For example, given the two strings $ACACT$ and $GACAC$, their Hamming distance is 5 since there are no matches, but the edit distance is 2 since it suffices to add G and remove T.

From these definitions, we can make the observation that the edit distance takes into account information about the ordering of characters and captures the best alignment between two strings. However, while Hamming distance has complexity that is linear with the string length, edit distance has complexity that is quadratic. While several dynamic programming optimizations exist for accelerating edit distance computations [20], they often rely on pre-specified distance thresholds and are unable to provide performance competitive with fine-tuned Hamming distance computations in practice. Given the complexity of this metric, we rely on randomized embedding techniques to minimize the overhead of the edit distance computations.

Randomized embedding refers to a set of methods that map a complex metric space into a simpler one. CGK-embedding algorithm, recently proposed by Chakraborthy et al. [4] is one such algorithm that can map problems from an edit space into a Hamming space. Given two strings x, y of length N taken from an alphabet $\sum$ such that $d_E(x, y)$, the edit distance between x and y, is less than K, CGK-embedding is a function $f\colon \sum^N \to \sum^{3N}$ that maps strings x and y into $f(x)$ and $f(y)$ such that, with probability at least 0.99, the Hamming distance of $d_H(f(x), f(y))$ is bounded by K^2 when $d_E(x, y) < K$. This implies that the distortion D, defined as the ratio $D(x, y) = \frac{d_H(f(x), f(y))}{d_E(x, y)}$, is at most K. Thus, as long as the edit distance is small, the distortion of embedding is small. This implies that the Hamming distance of embedded strings will accurately track the edit distance of the original strings, thereby making it possible to replace the expensive edit distance computation with cheap Hamming distance computation [22].

Algorithm 1. CGK-embedding

Input: A string $S \in \{A, C, G, T\}^N$, a random string $R \in \{0, 1\}^{3N}$ and a char for padding $P = 0$

Output: The embedded string $S' \in \sum^{3N}$

1: $i \leftarrow 0$
2: **for** $j = 0 \to 3N - 1$ **do**
3:      **if** $i < N$ **then**
4:          $S'_j \leftarrow S_i$
5:      **else**
6:          $S'_j \leftarrow P$
7:      **end if**
8:      $i \leftarrow i + R_j$
9: **end for**
10: **return** S'

The pseudo-code of embedding algorithm is shown in Algorithm 1. In this case the procedure is applied to all strings of length N that are composed of the characters A, C, G, T, representing the DNA alphabet. Given an input string, the algorithm builds the corresponding embedded representation by appending one character at time taken from the input string. The character appended can be the repetition of the previous character or the next character in the input string according to the value of a binary random string. In other words, the pointer of the current character in the input string increases or remains the same depending on the random string value, that can be 0 or 1. When the pointer to the input string goes beyond the string length, the embedded string is padded with a special character P. In general, P can be any character that is not included in the alphabet of the given dataset S. For sake of simplicity, in Algorithm 1 we use 0 for padding. In essence, what we get as the output of embedding is a string where characters from the input string can be repeated one or more times.

LSH for Hamming Distance. One of the main advantages of moving from the edit distance space to the Hamming space is being able to use some useful algorithms that are valid in the Hamming space only and instead not applicable to the edit distance. One of these algorithms is Locality Sensitive Hashing (LSH) [12].

Definition 1. *We call a family $\mathcal{H}$ of functions (d_1, d_2, p_1, p_2)-sensitive for a distance function D if for any $p, q \in U$ (where U is the item universe):*

- *if $D(p,q) \leq d_1$ then $\mathbb{P}[h(p) = h(q)] \geq p_1$, that is, if p and q are close, the probability of a hash collision is high;*
- *if $D(p,q) \geq d_2$ then $\mathbb{P}[h(p) = h(q)] \leq p_2$, that is, if p and q are far, the probability of a hash collision is low;*

where $h \in_r H$ are hash functions randomly sampled from the family of hash functions $\mathcal{H}$

Considering two bit-string p and q of length N. In the Hamming distance case, the hash function is defined as the ith bit of these strings. Thus, if their Hamming distance is $d_H(p,q)$, that is, the number of bits that differ position wise in the two strings, then the probability that any given bit at a random position is the same in both strings is $1 - \frac{d_H(p,q)}{N}$. Thus, the bit-sampling LSH family for Hamming distance, defined as: $H_N = \{h_i : h_i(b_1...b_N)) = b_i \mid i \in [N]\}$ is $(d_1, d_2, 1 - \frac{d_1}{N}, 1 - \frac{d_2}{N})$-sensitive for the two Hamming distances $d_1 < d_2$.

We use Hamming LSH over embedded reads to separate out the reads into different buckets such that with a very high probability, reads within a bucket are similar to each other, and hence correspond to the same reference.

Clustering Based on Edit Distance. At this stage, we have all reads grouped in hash buckets based on their similarity. However, we still have two problems to solve. (1) LSH can produce false positives, meaning that two dissimilar reads can end up in the same bucket. The main consequences is that if reads are very different, the consensus procedure lead to the wrong result. (2) Reads can have different lengths due to insertions and deletions errors. Thus, we need to adjust the reads in order to make their lengths uniform while taking into account possible insertion/deletion errors. Both these problems can be solved by aligning the reads in each bucket. More specifically, given a bucket, we sort the reads based on length such that reads with length matching the reference oligos are moved to the front of the bucket. Then, starting with the first read in the bucket, we align all the following reads to the first one. The intuition behind sorting the reads is as follows. If a read has the same length as the reference oligos (160nt), either the read has no insertion/deletion errors, or there are an even number of insertion and deletion errors. Since the probability of errors is low with short-read sequencing, the former scenario is more likely. Thus, by picking reads that are of the correct length and aligning the rest of the reads to it, we increase the probability of finding the correct original reference oligo.

The alignment of reads gives two pieces of information: the edit distance and the Compact Idiosyncratic Gapped Alignment Report (CIGAR). The edit distance allows us to identify reads that are actually dissimilar even if they are in the same buckets. We group only similar reads to form a cluster. The CIGAR contains the base-by-base alignment information (the sequence of matches, insertions and deletions) needed to align one read to the other. Using this information, we adjust the reads by adding gaps where there is a deletion error, or deleting nucleotides where there is an insertion error. Once all similar reads are found within the bucket, we save the cluster and remove the reads from the bucket. Any dissimilar reads that were not a part of the cluster are still left in the bucket. In order to deal with false positives by LSH, the remaining reads are then processed again in the same way, with the procedure being repeated until the bucket is empty.

Position-Wise Consensus. The result of the previous stages is a set of clusters. All reads within a cluster are noisy copies of the same reference oligo with some errors in random positions. At this point of the algorithm the only missing step is the consensus procedure. The consensus algorithm works on a per position basis. The key idea is that despite some random errors, all reads in a cluster are aligned and adjusted based on the edit distance and CIGAR. Thus, if we consider a specific position across all reads, it is likely that the majority of reads in that position will contain the correct nucleotide while only some of them report the wrong one. This implies that for any given position, it will be enough to take the most-frequent nucleotide as the consensus outcome. Repeating this procedure for each position, we produce one inferred oligo per cluster. We would like to point out here that not all oligos need to be correctly inferred. In fact, as we show later, some original oligos might not appear at all in the inferred set, and other inferred oligos might have errors. We rely on the parity added by LDPC codes at a higher level to recover data despite these errors.

The inferred oligos are then passed to the decoder which reverses the steps shown in Fig. 2. The constrained code is first used to convert each 160nt oligo back into 300-bit sequences by converting each individual 16-nt motif into its corresponding 30-bit value. The index stored in each 300 bits is used to reassemble bits back in the correct order. The LDPC decoder is then used to recover back data even if some bits were wrongly decoded, or some bits were zeroed out as corresponding oligos were missing. The decoded data is then derandomized to obtain a stream of bits that corresponds to the original input.

4   Evaluation

At this stage of our collaboration, we have assembled the entire pipeline. We are in the process of carrying out a real-life, large-scale synthesis experiment. As we do not have results from real experiments yet, we will provide a preliminary, simulation-driven evaluation in this section. Note that we simulate only the synthesis and sequencing steps in Fig. 1. We encode/decode the real dataset using our pipeline.

4.1   Experimental Setup

The raw SIARD archive that is fed as input to our pipeline is 12.9 MB in size. With redundancy added by LDPC, the resulting binary data to be stored on DNA is 15.19 MB in size. We encode the SIARD archive generating 405,212 oligos, each with a length of 160nts. Using these original oligos, we then generate four million reads by using a short-read simulator[3] tool, that adds random errors such as insertion, substitution, and deletion in each read to mimic the actions of an Illumina DNA sequencer. This corresponds to an average coverage of $10\times$, meaning that each oligo, on average is covered by 10 noisy copies.

4.2   Benchmark with Sequence Alignment

We begin our analysis by visualizing the coverage across oligos. While the average coverage is $10\times$, the overall coverage typically follows a negative binomial distribution, with some oligos being covered hundreds of times, and some being not covered at all. To visualize this, we aligned the reads to reference with BWA-MEM v0.7.17 [14], a state-of-the-art short-read aligner. Based on the alignment result from BWA-MEM, we show the histogram of coverage across oligos in Fig. 4. The x-axis is the coverage (or number of reads that map to an oligo), and y-axis is the number of oligos with that coverage in log scale. As can be seen, the coverage distribution spans a range from 1 to 26, with majority of oligos being covered 5–15× as expected given that the simulator was configured to produce $10\times$ coverage.

We also use the alignment result to show the histogram of number of errors (indels and substitutions) in Fig. 5, where x-axis represents the number of errors, and y-axis represents the number of oligos with that error count in log scale. As can be seen, the error distribution is right skewed, with a majority of reads having fewer than 2 mismatches, and a few reads having as many as 8 mismatches which accounts for a 5% error rate given the oligo length of 160. Finally, we also used the alignment result to verify that each read uniquely maps to an oligo (absence of XA' field in the SAM output file produced by BWA-MEM). This shows that oligos are "far" from each other in terms of edit distance due to randomization.

---

[3] https://sourceforge.net/projects/bbmap/.

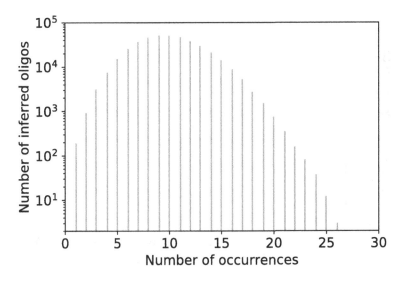

Fig. 4. Histogram of occurrence of inferred oligos

BWA-MEM performs base-by-base, edit-distance-based base-by-base align-ment and reports CIGAR and alignment score. As a result, it is computationally very intensive. While such an alignment is required for genomic data analysis in order to determine the exact location of a read in the genome, for the purpose of DNA storage, we found that it is sufficient to simply map each read to an oligo without full alignment. Accel-Align (v1.1.1; [21]) is a short-read aligner that we have developed in the context of project OligoArchive that supports alignment-free mapping-only mode that can quickly map reads to oligos. Thus, we present a comparative analysis of BWA-MEM and Accel-Align here with the goal of presenting it as a open-source tool that can be used by other researchers for both DNA storage and more broadly, for analyzing genomic data.

Table 1. Performance and accuracy of BWA-MEM and Accel-Align (map mode)

	Exec. time (second)	Correctly aligned (%)
BWA-MEM	25.5	100
Accel-Align	3.8	99.9

We use the BWA-MEM and Accel-Align to align/map reads to oligos. Using BWA-MEM's alignment as the gold standard, we evaluate Accel-Align's accu-racy. Table 1 shows the execution time and the percent of correctly aligned reads. It shows that Accel-Align can performing mapping 6x faster than BWA-MEM at a slight drop in accuracy of 0.004%. On further analysis, we found this drop to

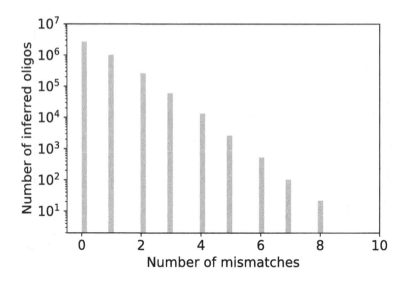

Fig. 5. Histogram of mismatches

be due to Accel-Align's inability to map reads with very high error rate. Accel-Align is specifically built to trade off performance and accuracy for Illumina-based short-read sequencing reads whose error rate is typically much lower than 5%. As our simulated reads offer a more pessimistic error model, Accel-Align experiences a slight drop in accuracy. Despite this, as the coverage histogram with Accel-Align and BWA-MEM are near identical, we have found Accel-Align to be a very useful tool for analyzing both DNA storage reads and more broadly, genomic data [21].

4.3   End-to-End Decoding Results

Having presented an analysis of the reads, we will now present the decoding results. Using OneConsensus procedure described earlier, we obtain the inferred oligos using the simulated dataset. Table 2 shows the error statistics for these oligos. The figures are obtained by comparing the inferred oligo for a certain index with the corresponding original reference oligo. We see that we are able to infer 404,075 oligos that correspond to 99.7% of the original oligos perfectly without errors. In addition, 1010 oligos were inferred with some errors and 127 oligos were completely missing. Note that errors in an oligo does not imply that the entire oligo is different from the original, but differs only with respect to a few motifs. For this reason, we also report the difference between inferred data and original encoded file in terms of number of bits.

We then use the constrained code to convert these inferred oligos into 300-bit sequences, and reassemble them in order based on the 19-bit index. At this stage, we will certainly have situations where an oligo is missing due to sequencing simulation bias, or an oligo could not be converted back into 300-bits due to errors. In both cases, there will be a corresponding index whose data bits cannot be recovered. We insert a sequence of zero bits for such indices and use the reconstructed binary together with the LDPC decoder to restore the original input. Despite the errors reported in Table 2, the LDPC decoder was able to recover back the original archive completely, thanks to the additional parity information added during encoding.

Table 2. Statistics for decoding of SIARD archive

#Original Oligos	405212
#Correctly Inferred Oligos	404075
#Incorrectly Inferred Oligos	1010
#Missing Oligos	127
#Incorrect bits	42678

Finally, in this section we compare OneConsensus with OneJoin-based consensus algorithm. Table 3 compares the two algorithms in terms of the accuracy achieved in inferring the encoding oligos. In terms of the number of oligos correctly inferred in their entirety, meaning an exact match between the inferred oligo and the original reference oligo, we see that OneConsensus slightly underforms OneJoin. But as we mentioned earlier, oligos can also differ by just a few motifs only, making the statistic about the correctly inferred oligos insufficient to determine the actual accuracy of the two algorithms. For this reason, we report also the number of missing oligos and the number of bits wrongly inferred by the two algorithms. The number shows that OneConsensus outperforms OneJoin, as it mistakes only misses 127 oligos, while OneJoin 595 oligos. Overall, OneConsensus leads to 42678 bit errors, compared to the 95935 bit errors produced by the OneJoin-based consensus.

In terms of memory consumption, we observed that OneJoin reaches a peak of 2.5 GB, while OneConsensus requires only 1.1 GB. While the difference may not seem too striking for this dataset, we would like to point out that while OneConsensus memory consumption grows linearly with the dataset size, One-Join requires an amount of memory that is quadratic with coverage.

Table 3. Statistics for OneConsensus and OneJoin-based consensus

	OneConsensus	OneJoin
#Correctly Inferred Oligos	404075	404104
#Incorrectly Inferred Oligos	1010	513
#Missing Oligos	127	595
#Incorrect bits	42678	95935

5   Conclusion and Future Work

In this work, we provided an overview of the ongoing collaboration between project OligoArchive and the Danish National Archive in using DNA to preserve culturally significant digital data. Building on prior work on molecular information storage and digital preservation, we presented a holistic, end-to-end pipeline for preserving both data and the meaning of data on DNA, and tested the pipeline using simulation studies. There are several avenues of future work we are pursuing. First, as described earlier, we are in the process of carrying out a large-scale experiment to validate our pipeline using real data. Second, we are investigating various optimizations to both encoding and consensus algorithms to support alternate synthesis and sequencing technologies with potentially higher error rates. Finally, while we addressed the question of preserving data in this work, we left open the question of preserving the decoding algorithm itself. Recent work has investigated the design of nested universal emulators that can be used to preserve and emulate such decoders using analog media like film or archival paper [3]. Thus, we are investigating methods to combine DNA-based digital data storage with analog media-based decoding logic storage.

Acknowledgment. This work was partially funded by the European Union's Horizon 2020 research and innovation programme, project OligoArchive (grant agreement No. 863320).

References

1. Digital Preservation Handbook. Digital Preservation Coalition (2015)
2. Appuswamy, R., et al.: OligoArchive: using DNA in the DBMS storage hierarchy. In: CIDR (2019)
3. Appuswamy, R., Joguin, V.: Universal layout emulation for long-term database archival. In: CIDR (2021)
4. Chakraborty, D., Goldenberg, E., Koucký, M.: Streaming algorithms for embedding and computing edit distance in the low distance regime. In: Proceedings of the Forty-Eighth Annual ACM Symposium on Theory of Computing, pp. 712–725 (2016)
5. Chandak, S., et al.: Improved read/write cost tradeoff in DNA-based data storage using LDPC codes. In: 2019 57th Annual Allerton Conference on Communication, Control, and Computing (2019)
6. Church, G.M., Gao, Y., Kosuri, S.: Next-generation digital information storage in DNA. Science **337**(6102), 1628–1628 (2012)
7. of Congress, L.: SIARD (Software Independent Archiving of Relational Databases) Version 1.0 (2015). www.loc.gov/preservation/digital/formats/fdd/fdd000426. shtml. Accessed 28 May 2021
8. Corporation, S.R.: 2018 semiconductor synthetic biology roadmap. https://www. src.org/program/grc/semisynbio/ssb-roadmap-2018-1st-edition_e1004.pdf (2018)
9. Erlich, Y., Zielinski, D.: DNA Fountain enables a robust and efficient storage architecture. Science **355**(6328), 950–954 (2017)

10. Fontana, R.E., Decad, G.M.: Mooreâs law realities for recording systems and memory storage components: Hdd, tape, nand, and optical. AIP Adv. **8**(5), 056506 (2018)
11. Gallager, R.: Low-density parity-check codes. IRE Trans. Inf. Theory **8**(1), 21–28 (1962)
12. Gionis, A., Indyk, P., Motwani, R.: Similarity search in high dimensions via hashing. In: Proceedings of the 25th International Conference on Very Large Data Bases, pp. 518–529. VLDB 19999 (1999)
13. Goldman, N., et al.: Toward Practical High-capacity Low-maintenance Storage of Digital Information in Synthesised DNA. Nature **494**, 77–80 (2013)
14. Li, H.: Aligning sequence reads, clone sequences and assembly contigs with bwa-mem. arXiv preprint arXiv:1303.3997 (2013)
15. Marinelli, E., Appuswamy, R.: Onejoin: cross-architecture, scalable edit similarity join for DNA data storage using oneapi. In: ADMS (2021)
16. Organick, L., et al.: Random access in large-scale DNA data storage. Nat. Methods **11**(5) (2014)
17. Perlmutter, M.: The lost picture show. https://tinyurl.com/y9woh4e3 (2017)
18. Shapiro, B.: Mammoth 2.0: will genome engineering resurrect extinct species? Genome Biol. **16**, 1–3 (2015)
19. SNIA: 100 year archive requirements survey 10 years later. https://tinyurl.com/yytsbvmb (2017)
20. Ukkonen, E.: Algorithms for approximate string matching. Inf. Control **64**(1), 100–118 (1985)
21. Yan, Y., Chaturvedi, N., Appuswamy, R.: Accel-align: a fast sequence mapper and aligner based on the seed-embed-extend method. BMC Bioinform. **22**, 1–20 (2021)
22. Zhang, H., Zhang, Q.: Embedjoin: efficient edit similarity joins via embeddings. In: Proceedings of the 23rd ACM SIGKDD International Conference on Knowledge Discovery and Data Mining, pp. 585–594 (2017)

Author Index

Printed in the United States
by Baker & Taylor Publisher Services